Introduction to Numerical Methods for Water Resources

Introduction to Numerical Methods for Water Resources

W. L. WOOD

CLARENDON PRESS·OXFORD
1993

Oxford University Press, Walton Street, Oxford OX2 6DP
Oxford New York Toronto
Delhi Bombay Calcutta Madras Karachi
Kuala Lumpur Singapore Hong Kong Tokyo
Nairobi Dar es Salaam Cape Town
Melbourne Auckland Madrid
and associated companies in
Berlin Ibadan

Oxford is a trade mark of Oxford University Press

Published in the United States
by Oxford University Press Inc., New York

A catalogue record for this book is available from the British Library

Library of Congress Cataloging in Publication Data
Wood, W. L.
Introduction to numerical methods for water resources / W. L. Wood.
Includes bibliographical references and index
1. Hydraulics—Statistical methods.   I. Title.
TC163.W66 1993   628.1—dc20   92-32900

ISBN 0-19-859690-1

Set by Integral Typesetting Ltd, Great Yarmouth, Norfolk
Printed in Great Britain on acid-free paper by
Biddles Ltd., Guildford & King's Lynn

Preface

Numerical methods provide a powerful and essential tool for the solution of problems of water resources. Finite difference methods, finite element methods, the integrated finite difference method, the method of characteristics, boundary integral methods, finite volume, or cell-differencing methods—these are all in current use for the numerical solution of problems of water supply and management. No one of these has been shown to be superior to all the others for the solution of surface and subsurface flow and for water-quality modelling. Different methods work well in different situations and some problems require combinations of several methods. It is essential to know something about all of them in order to make a reasoned judgement of the merits of current practices.

The inspiration to write this book has come from my work as consultant to the Institute of Hydrology and Halcrow Water and collaboration with a number of people in the water industry arising from student projects. It is meant to help water engineers, hydrologists, geohydrologists, and applied mathematicians working with them to understand the mathematics which are the basis of the numerical codes now easily available. There is an elementary introduction to the standard methods listed above and a description of their use for surface and subsurface flow for water-quality modelling. Outlines of more specialized versions are given with many references. I assume some knowledge of elementary numerical analysis; references are given to Conte and de Boor (1980), and to Spencer *et al.* (1977), for background.

This book has evolved from my collaboration with many people. I should particularly like to thank John Barker, Ann Calver, and Gareth Pender for reading the first draft and making useful comments. Thanks are also due to James Bathurst, David Cooper, Roger Falconer, Rob Hiley, George Mitchell, Paul Samuels, Dominic Reeve, Anders Refsgaard, and Ewan Wilson for helpful correspondence and copies of papers, to Robert Muir Wood for expert help with the figures, and to Alan Muir Wood for general assistance.

Department of Mathematics   W.L.W.
University of Reading
March 1992

Acknowledgements

Figures and tables from papers jointly by the author and others are identified in the text. These are reprinted by permission as follows:

from the *International Journal of Numerical Methods in Engineering*, the *International Journal of Numerical Methods in Fluids*, and *Communications in Applied Numerical Methods* by permission of John Wiley and Sons Ltd;

from the *Journal of Hydrology* by permission of Elsevier Science Publishers BV;

from the *Proceedings of the Institution of Civil Engineers* by permission of Thomas Telford Publications.

Acknowledgement is also due to Peter Arnold, Brian Connorton, Andrew Griffiths, and Roy Wikramaratna for permission to use material from their M.Sc. dissertations, to Thames Water Utilities for permission to use Figs 1.1.1 and 7.3.1, and to Sir William Halcrow and Partners for permission to use Figs 7.3.2 and 7.4.1.

Contents

4. Numerical solution of surface flow equations by finite differences

Notation

This list contains the definitions of the symbols in this book and the sections where they are first used. All these symbols are defined in the text. There is some duplication but it is hoped that this will not cause any confusion.

Symbol	Definition	Dimensions	Section
a	a length	L	8.3
a_L, a_T	longitudinal and transverse dispersivities	L	9.1
$a_m(t)$	amplitude in Fourier mode m		3.7
a, b	parameters in box scheme		3.9
a, b, c	parameters in a formula		6.2
a, b, f	parameters in hyperbolic equations		2.1
a_j, b_j, c_j	parameters in difference equations		3.5
$[a, b]$	one-dimensional region		3.5
A	wetted cross-sectional area of channel	L^2	1.2
A	parameter in Fourier analysis		3.10
$\mathbf{A}$	matrix of simultaneous linear equations		3.5
A_1, A_2	shorthand parameters	L^{-2}	5.4
A, B	constants of integration	L^2	1.8
A, B	parameters in solution of characteristic polynomial		3.3
A, B, C	finite element triangle		6.2
A_c	area of overbank flow	L^2	4.4
b	constant breadth of channel	L	1.4
b_1	parameter in finite element	L^2	6.2
B	constant depth of aquifer	L	1.7
B	dimensionless parameter in stream leakage		5.5
B	hillslope width	L	7.7
$\mathbf{B}$	magnetic field		8.3
B_c	surface width of overbank flow	L	4.4
c	kinematic wave speed	LT^{-1}	1.5
c	constant average velocity in channel	LT^{-1}	1.6

$\hat{c} = \mathrm{d}q/\mathrm{d}A$		LT^{-1}	1.5
c_1, d_1	parameters in finite element	L	6.2
$C(\psi)$	specific water storativity	L^{-1}	7.7
C_b	parameter in overbank flow		4.3
C_c	parameter in optimum time step formula		4.3
c_n	averaged value of $c(q)$	LT^{-1}	4.2
C	concentration of solute		1.1
C_j	concentration of solute j		9.1
C_L	source concentration of pollutant		9.2
C_1	maximum concentration of salt water		9.5
$\mathbf{C_N}$	Courant number		3.9
C'	Chezy parameter	$L^{1/2}T$	1.2
d	representative length	L	1.7
d	depth of saturated region	L	7.9
d_P	height of point P on D–F parabola	L	7.9
d_a	adjusted value of d_P	L	7.9
D	coefficient of diffusion	L^2T^{-1}	1.6
D^*	coefficient of molecular diffusion	L^2T^{-1}	9.1
D_{ij}	coefficients in diffusion tensor	L^2T^{-1}	9.2
$D_{j,k}, W_{j,k}, E_{j,k}, N_{j,k}, S_{j,k}$	shorthand notation	T^{-1}	5.2
D_{xx}, etc.	depth-averaged diffusion coefficients	L^2T^{-1}	9.2
e_n	error in finite difference solution		3.4
e_j^n	error in finite difference solution		3.10
e_{max}	tolerance in Cooley algorithm	L	7.7
E	a large number		6.2
f	friction parameter		1.2
$f(x), f(x, y)$	external force effects		3.6
$f(y, t)$	law of decay of effluent	T^{-1}	3.2
$f(x, t)$	representative function		4.3
f_0	initial value of $f(y, t)$	T^{-1}	3.2
f_i	component of vector $\mathbf{F}$	L^3T^{-1}	6.2
$f_n = f(y_n, n\Delta t)$		T^{-1}	3.2
$f(x)$	function in differential equation		3.5
f, g	parameters in method of characteristics		2.4
F	friction force per unit area for surface flow	$ML^{-1}T^{-2}$	1.2
$\mathbf{F}$	Froude number		4.3

F	'force vector', vector of known values	L^3T^{-1}	6.2
F	decay coefficient	T^{-1}	9.2
$\bar{F}$	average value of F over a time step	L^3T^{-1}	7.2
g	acceleration due to gravity	LT^{-2}	1.2
g	parameter in example		3.6
$g(x,y)$	known head	L	6.1
G, G_b	denote physical properties		A.1
h	height of channel water surface	L	1.4
h_b	height of channel base	L	9.2
h	size of finite element	L	6.1
h, h_b	heights of water table and base of aquifer	L	1.7
h_f	height of phreatic surface above sea level	L	9.5
h_s	depth of salt/fresh interface below sea level	L	9.5
$\mathbf{h}_n$	vector for nodal values at time step n	L	7.2
h^k	kth iterate for h	L	7.2
$\mathbf{h}_n^k$	kth iterate for $\mathbf{h}_n$	L	7.2
$\hat{\mathbf{h}}^k$	adjusted kth iterate	L	7.2
$\hat{h}_j$	finite element solution at node j	L	7.5
h_w	height of water table at a well	L	1.8
h_0	height of water table where there is zero drawdown effect	L	1.8
H	vector in solution of Richards' equation	L^2T^{-2}	7.7
$H = h - h_b$	depth of surface water	L	1.3
H_0	depth at top of slope	L	2.2
H_1, H_2	parameters in method of characteristics		2.3
j, k	grid point numbers		3.6
J	number of basis functions		6.1
$\mathbf{J}$	Jacobian matrix		6.6
$\mathbf{J} = (J_1, J_2, J_3)^T$	mass flux vector	$ML^{-2}T^{-1}$	1.7
k	constant of proportionality for the friction force		1.2
k	kinematic wavenumber		10.1
k_{ij}	(i,j)th entry in 'stiffness matrix' K	L^2T^{-1}	6.2

$k_{ij}^{(e)}$	(i, j)th entry in element stiffness matrix	L^2T^{-1}	6.2
k_{ij}	permeability tensor	L^2	9.5
$\mathbf{K}$	'stiffness matrix'	L^2T^{-1}	6.2
$\mathbf{K}^{(e)}$	element stiffness matrix	L^2T^{-1}	6.2
K	hydraulic conductivity	LT^{-1}	1.7
K_{ij}	block entries in stiffness matrix		8.4
K_{ij}	hydraulic conductivity tensor	LT^{-1}	9.3
K_x, K_y, K_z	components of hydraulic conductivity	LT^{-1}	7.7
K_s	saturated hydraulic conductivity	LT^{-1}	7.7
$\bar{K}$	vertically averaged hydraulic conductivity	LT^{-1}	1.8
$K_{sl}(x, y)$	stream leakage parameter	T^{-1}	5.5
$\bar{K}_x, \bar{K}_y$	components of the vertically averaged hydraulic conductivity	LT^{-1}	1.7
l	depth of soil in hillslope	L	7.9
L	representative length	L	1.2
$L_1, L_2, \ldots$	area coordinates		6.2
$L_{\Delta t}(\cdot), L_{\Delta x}(\cdot),$	difference operators		3.2
$L_{\Delta x, \Delta y}(\cdot)$			3.6
m	mode number		3.7
m_{ij}	(i, j)th entry in mass matrix	L^2	6.2
$m_{ij}^{(e)}$	(i, j)th entry in element mass matrix	L^2	6.2
m, n	powers in friction law		1.2
M	mass of element in channel flow	M	1.4
$\mathbf{M}$	'mass matrix' in the finite element method	L^2	6.2
n	porosity of ground		1.7
n	number of time steps		3.2
$\mathbf{n}$	unit outward normal		3.6
n_x, n_y	components of unit outward normal		6.1
$\hat{\mathbf{n}}_b$	unit normal to base of aquifer		A.1
$\mathbf{N}$	vector of surface rate of accretion per unit area	LT^{-1}	A.1
N	vertical components of surface rate of accretion	LT^{-1}	A.1
N_n	number of nodes		7.8
N_t	number of time steps		7.8

Symbol	Description	Dimensions	Section
$N_j(x, y)$	basis functions for the finite element method		6.1
p	mode number		3.9
p	pressure	$ML^{-1}T^{-2}$	9.5
$p, q,$ $p(x), q(x)$	parameters in differential equation examples		3.5
p, q	parameters in Fourier mode		5.4
p_j, q_j	values of $p(x), q(x)$ at $x = j\Delta x$		3.5
P	wetted perimeter of channel	L	1.2
P_1, P_2	pressures	$ML^{-1}T^{-2}$	1.4
$\mathbf{P}_e$	Peclet number		3.8
P, Q	parameters in kinematic wave equation		1.5
q	discharge in channel flow	L^3T^{-1}	1.4
Q_L	source per unit length in channel flow	L^2T^{-1}	9.2
$\mathbf{q}$	specific discharge in groundwater flow	LT^{-1}	1.7
$\hat{\mathbf{q}}$	vector of nodal values of q	LT^{-1}	7.5
q_b	interchange between channel and flood plain	L^3T^{-1}	4.4
q_j^n	approximate value of $q(j\Delta x, n\Delta t)$	L^3T^{-1}	3.8
$\bar{q}^n = (q_j^n + q_{j+1}^n)/2$		L^3T^{-1}	3.8
q_x, q_y, q_z	components of specific discharge	LT^{-1}	A.1
Q	source of sink in aquifer	LT^{-1}	1.7
Q	discharge per unit width	L^2T^{-1}	10.2
$\tilde{Q}$	source or sink in three-dimensional groundwater problem	T^{-1}	7.7
$[Q]$	net inflow from wells in aquifer	L^3T^{-1}	5.2
Q_b	baseflow per unit width	L^2T^{-1}	10.2
Q_p	peak discharge	L^3T^{-1}	4.2
Q_s	stream leakage	LT^{-1}	5.5
Q_w	abstraction rate for well	L^3T^{-1}	1.8
Q_w	(dimensionless) strength of source		8.2
$\hat{Q}_n$	abstraction rate from well n	LT^{-1}	5.2
Q, Q_k	discharge at foot of slope	L^3T^{-1}	7.8
Q_k^b	baseline discharge at foot of slope	L^3T^{-1}	7.8

r	rate of rainfall	LT^{-1}	1.4
$r(x)$	lateral inflow to channel	LT^{-1}	2.3
r	polar coordinate	L	1.8
r, r_1, r_2	roots of characteristic polynomial		3.3
r_w	radius of well	L	1.8
$\hat{r}$	distance from singularity	L	8.1
r_0	(dimensionless) radius of σ		8.1
r_{ij}	(dimensionless) distance $P_i P_j$		8.1
r_w	(dimensionless) distance from source		8.2
r_t	radius of tunnel	L	8.3
R	hydraulic radius in channel	L	1.2
R	radial distance from well	L	1.8
R	groundwater region	L^2	3.6
R_i, R_o, R_r, R_g	groundwater regions		8.3
			8.4
$\mathbf{R_e}$	Reynolds' number		1.7
s	slope		1.2
s	parameter in Cooley algorithm		7.7
s_x, s_z	slopes in x, z directions		1.4
$s(r, t)$	well drawdown	L	1.8
s_f	friction slope		1.2
s_{fx}, s_{fz}	friction slopes in x, z directions		1.4
S	effective storativity (in two dimensions)		1.7
$\bar{S}$	local average value of effective storativity		7.2
S_0	specific storativity (in three dimensions)	L^{-1}	1.7
S	boundary of region R	L	3.6
S_i, S_o	boundaries of R_i, R_o		8.3
S_1, S_2	known-head, known-flow boundaries	L	5.2
S_{jk}	coupling terms in chemical reactions	T^{-1}	9.1
S_{ir}, S_{ig}, S_{rg}	interfaces		8.4
t	time	T	1.4
t_n	time at which solution calculated	T	7.7
t_p	time to peak discharge	T	4.2
t_r	duration of rainfall	T	10.2

$T = n\Delta t$	a fixed time	T	3.4
T	transmissivity	L^2T^{-1}	1.7
$\bar{T}$	average value of T over element	L^2T^{-1}	7.2
T_x, T_y	components of transmissivity	L^2T^{-1}	1.7
u	depth-averaged velocity	LT^{-1}	1.3
u, v, w	components of velocity in surface flow	LT^{-1}	1.4
$u(x)$	velocity in channel cross-section	LT^{-1}	3.5
$u(x, t),$ $u(x, y, t)$	representative functions in parabolic problem examples		3.7
u_j	approximate value of $u(j\Delta x)$		3.5
u_j^n	approximate value of $u(j\Delta x, n\Delta t)$		3.7
$\mathbf{u}$	vector of approximate values at grid points		3.5
$\hat{\mathbf{u}}$	vector of exact values at grid points		3.5
u	label for $\psi_{\text{in}} = $ constant		7.9
U, V	functions in example		8.1
v	average velocity over cross-section of channel	LT^{-1}	3.8
$v_\mathrm{P}, v_\mathrm{Q}$	values of v at points P, Q	LT^{-1}	2.4
$\mathbf{v}$	velocity of 'particles carrying the property G'	LT^{-1}	A.1
v	label for $\psi_{\text{in}} = 0.1z$		7.9
V	amount of effluent	L	1.6
$\mathbf{V}$	Vedernikov number		4.5
w	weighting parameter		3.8
$w(x, y)$	weight or test function in the finite element method		6.1
w_j	weighting parameter		6.6
x	horizontal distance across channel	L	1.3
x	distance down slope	L	1.4
x_j	coordinate in jth direction	L	9.5
x, y, z	coordinates: x, y horizontal, z vertical	L	1.7
(x_j, y_j)	nodal coordinates in the finite element method	L	6.4
$(\bar{x}, \bar{y})$	centroid of triangle	L	6.6
$y_\mathrm{P}, y_\mathrm{Q}$	values of y at P, Q	L	2.4
y_n	approximate value of $y(n\Delta t)$		3.2

y_j^n	approximate value of $y(j\Delta x, n\Delta t)$	L	3.8
$y(0)$	initial value of $y(t)$		3.2
y_{n+1}^k	kth estimate of y_{n+1}		3.2
α, β	parameters in friction laws		1.5
α, β, γ	parameters in boundary conditions		3.5
$\alpha_{p,q}$	amplitude in Fourier mode	L	5.4
β	momentum correction coefficient		1.4
$\beta(x, y)$	known flow	L^2T^{-1}	6.1
β_n	parameter in convergence analysis		3.4
$\beta_n, \varepsilon_n,$ $\beta_{n,p}, \varepsilon_{n,p}$	parameters in error terms		3.10
$\delta(x-x_0, y-y_0)$	Dirac delta generalized function	L^{-2}	5.2
δ_{ij}	Kronecker delta		9.1
Δ_e	area of triangle element	L^2	6.2
$\Delta s, \Delta n$	dispersive spreading 'lengths'	$L^{1/2}T^{1/2}$	9.2
Δt	increment in t	T	2.3
Δt_n	increment in t at nth time step	T	7.7
Δ_k	parameter in Cooley algorithm	L	7.7
$\Delta x, \Delta y$	increments in x, y	L	2.3
ε	turbulence exchange parameter	L^2T^{-1}	1.3
ε_1	principal part of error		2.3
η	dummy variable in integral		1.6
η	(dimensionless) length		8.1
θ	slope angle		1.2
θ	parameter in θ method		3.3
θ	soil moisture content by volume		7.7
θ_s	saturated value of θ		7.7
θ, θ'	parameters		6.5
λ	parameter in method of characteristics		2.3
λ	parameter in various examples		3.3
$\lambda = c\Delta t/\Delta x$			3.9
μ	viscosity	$ML^{-1}T^{-1}$	10.1
v	kinematic viscosity	L^2T^{-1}	1.7
ξ, ξ_m	amplification factors		3.3
ξ, η	local coordinates		6.3

$\pi_1, \pi_2, \ldots$	dimensionless parameters		10.2
ρ	density	ML^{-3}	1.2
ρ_f, ρ_s	densities of fresh and salt water	ML^{-3}	9.5
σ	coefficient of diffusion	L^2T^{-1}	3.7
σ	circle		8.1
$\sigma, \sigma_r, \sigma_g$	(dimensionless) conductivity		8.4
τ_n, τ_j, τ_j^n	local truncation errors		3.2
ϕ	piezometric head (hydraulic potential)	L	1.7
ϕ	(dimensionless) potential		8.4
ϕ_∞	potential valid at infinite distance	L	8.3
$\hat{\phi}$	solution of Laplace's equation	L	8.3
ϕ_i	potential in R_i	L	8.3
$\phi_j \ldots$	nodal values of $\phi^h(x, y)$	L	6.1
$\phi_j(t)$	nodal values of $\phi^h(x, y, t)$	L	6.1
$\phi^h(x, y),$ $\phi^h(x, y, t)$	finite element approximations	L	6.1
$\bar{\phi}$	average value of ϕ over time step	L	7.7
$\boldsymbol{\phi}_n$	vector of nodal values at time step n	L	7.7
$\phi_{i,n}$	approximate value of ϕ at node i at time step n	L	7.7
ψ	$\tan \psi$ is the slope of tangent to curve		3.2
ψ	angle subtended by tangents		8.1
ψ	(moisture) pressure potential	L	7.7
ψ'	dimensionless stream function		9.5
ψ_{in}	initial value of pressure potential	L	7.9
ω	weighting parameter		7.2
$\hat{\omega}, \omega_k$	relaxation parameters		7.7

$\nabla \phi \equiv \text{grad } \phi \equiv$ gradient of a scalar function $\phi(x, y, z)$

$$\equiv \begin{bmatrix} \partial\phi/\partial x \\ \partial\phi/\partial y \\ \partial\phi/\partial z \end{bmatrix} \text{ in three dimensions or } \begin{bmatrix} \partial\phi/\partial x \\ \partial\phi/\partial y \end{bmatrix} \text{ in two dimensions.}$$

$$\mathbf{V} \cdot \mathbf{v} \equiv \mathrm{div}\ \mathbf{v} \equiv \text{divergence of a vector function } \mathbf{v}, \begin{bmatrix} u \\ v \\ w \end{bmatrix},$$

$$\equiv \frac{\partial u}{\partial x} + \frac{\partial v}{\partial y} + \frac{\partial w}{\partial z} \text{ in three dimensions}$$

or with $\mathbf{v} = \begin{bmatrix} u \\ v \end{bmatrix}$, $\mathbf{V} \cdot \mathbf{v} \equiv \dfrac{\partial u}{\partial x} + \dfrac{\partial v}{\partial y}$ in two dimensions.

$$\mathbf{V}^2 \phi \equiv \mathrm{div}\ \mathrm{grad}\ \phi$$

$$\equiv \frac{\partial^2 \phi}{\partial x^2} + \frac{\partial^2 \phi}{\partial y^2} + \frac{\partial^2 \phi}{\partial z^2} \text{ in three dimensions}$$

or $\dfrac{\partial^2 \phi}{\partial x^2} + \dfrac{\partial^2 \phi}{\partial y^2}$ in two dimensions.

1 Surface and groundwater flow—types of equation

1.1 Introduction

The movements of groundwater and of surface water (in lakes, reservoirs, and rivers) are closely related. When the water level in a river is higher than the water level in the adjacent aquifer and the river bed is permeable, the water will flow from the river to feed the aquifer; falling groundwater levels in the vicinity of a spring can cause it to dry up and the river may disappear in that neighbourhood. When the water level in the aquifer is higher the river takes water away from the aquifer. Figure 1.1.1 from the Thames Water Lower Colne Gravels study illustrates these effects. Natural recharge and stream leakage from the Colne Brook combine to add to the groundwater and the River Thames is draining the water away.

Increasing water quality is assuming comparable importance to that of water quantity. Putting more fertilizer on the ground than the crops can take up means that the surplus is there to be washed into the ground with the rain. Minerals from colliery waste can also enter the ground from the surface. In some parts of the world the domestic water supply comes mainly from groundwater through abstraction wells. In other parts the water supply comes from reservoirs which are fed by rivers which are themselves fed from groundwater. It is therefore

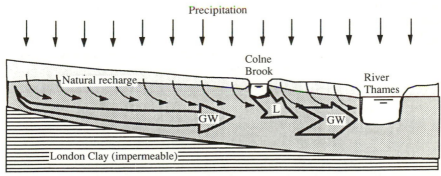

GW – groundwater flow
L   – leakage from stream to gravels

Figure 1.1.1: Surface and subsurface flow.

Table 1.1.1

	Initial value problems	Boundary value problems
Ordinary differential equations	$\dfrac{dC}{dt} = f(C, t),\ t > 0$ $C(0)$ given	$\dfrac{d}{dx}\left(p(x)\dfrac{du}{dx}\right) = g(u, x),\ a \le x \le b$ $u(a),\ u(b)$ given
Partial differential equations	(a) hyperbolic: $\dfrac{\partial H}{\partial t} = c\dfrac{\partial H}{\partial x},\ t > 0,\ x > 0$ $H(x, 0),\ H(0, t)$ given (b) parabolic: $\dfrac{\partial C}{\partial t} = \sigma\dfrac{\partial^2 C}{\partial x^2},\ t > 0,\ a \le x \le b$ $C(a, t),\ C(b, t),\ C(x, 0)$ given	elliptic: $\dfrac{\partial}{\partial x}\left(T_1\dfrac{\partial h}{\partial x}\right) + \dfrac{\partial}{\partial y}\left(T_2\dfrac{\partial h}{\partial y}\right) = Q$ in a region R in (x, y) plane with boundary conditions all round

important to have information about the passage of water and pollutants through the ground.

Thus the management of water resources must include consideration of both surface and groundwater flow. This book aims to give an elementary explanation of the mathematics needed to understand the simplest numerical models of surface and subsurface flow. These are the models most likely to be met in water resources computer packages and this chapter introduces the types of equation involved. Table 1.1.1 gives a summary of these types. The examples there are simplified versions of the equations whose numerical solutions are explained in this book. They include the equations representing surface and ground-water flow and pollutant transport:

1. The 'ordinary differential equation initial value problem' represents the decay with time of an effluent concentration $C = C(t)$. The initial value $C(0)$ of the concentration must be known.

2. The 'ordinary differential equation boundary value problem' repre-sents the lateral distribution of velocity u across a channel (section 1.3). This includes a second derivative and must have two boundary conditions.

3. The 'partial differential hyperbolic equation' is a simplified version of the kinematic wave equation, frequently used for surface flow (section 1.5). This has only first derivatives and needs the initial value $H(x, 0)$ and the value $H(0, t)$ at $x = 0$.

4. The 'partial differential parabolic equation' represents the diffusion with time of an effluent concentration C in a channel (section 1.6). (The symbol c has to be reserved for the celerity in the kinematic wave equation.) The effluent is assumed here to be dissolved in the water and its concentration is measured, for example, in 'parts per thousand', i.e. C is dimensionless. The initial distribution $C(x, 0)$ of the concentration must be known. Since the equation has a second derivative there must be two boundary conditions on the spatial variation. These are usually one at each end of the region, i.e. $C(a, t)$ and $C(b, t)$ are known.

5. The 'partial differential elliptic boundary value problem' represents the two dimensions in plan aquifer equation for the height h of the water in steady state groundwater flow (section 1.7). This includes second derivatives with reference to the space variables x, y and it is necessary to have boundary conditions all round the boundary, either known height or known flow. When the groundwater flow is changing with time the equation representing it becomes parabolic as shown in section 1.7, i.e. it has a first derivatives in time and second derivatives in the space coordinates. It is then necessary to have an initial condition as well as boundary conditions all round the boundary.

All the problems represented by differential equations in this book are assumed to be 'well posed'. This means that the equations are formulated so that there is a unique solution and small changes in the data make small changes in this solution; this is equivalent to the sensitivity analysis which should always be applied.

1.2 Surface flow—friction laws

The friction laws which give the terms used to represent the effect of surface roughness on surface flow are based originally on the assumption of uniform flow. This occurs with a long straight channel where the depth, cross-sectional area, and velocity in each cross-section are constant. The force of gravity is assumed to be exactly balanced by the resistance due to the surface roughness. Hence if the density is ρ, cross-sectional area A, control length L, slope angle θ given by

$\sin \theta = s$, assumed small, then balancing forces gives

$$\rho g A L s = P L F \tag{1.2.1}$$

where P is the wetted perimeter and F the friction force per unit area. If F is assumed to be proportional to the square of the average velocity v in the cross-section, this gives

$$\rho g A s = P k v^2 \tag{1.2.2}$$

where k is the constant of proportionality. Substituting $A/P = R$, the hydraulic radius, eqn (1.2.2) gives

$$s = C' \frac{v^2}{R} \tag{1.2.3}$$

where $C' = k/(\rho g)$.

This is generalized for non-uniform flow to a formula for the friction slope:

$$s_f = C'v^n/R^m. \tag{1.2.4}$$

Ackers (1958) gives values of n according to the type of flow as:

(1) $n = 1$ for laminar flow;

(2) $n = 1.75$ for smooth turbulent flow

(3) $1.74 \leq n \leq 2$ for transitional turbulent flow;

(4) $n = 2$ for fully turbulent flow.

$$\left. \right\} \tag{1.2.5}$$

Popular formulae for s_f are:

1. Manning: $n = 2$, $m = 4/3$, i.e.

$$s_f = C'v^2/R^{4/3}. \tag{1.2.6}$$

This is applied to pipes and open channels.

2. Chezy: $n = 2$, $m = 1$, i.e.

$$s_f = C'v^2/R. \tag{1.2.7}$$

This is also used in pipes and channels.

3. Darcy–Weisbach also uses $n = 2$ and $m = 1$ in the form

$$s_f = \frac{f}{8g} \frac{v^2}{R}. \tag{1.2.8}$$

In these three formulae suitable values of the parameters C' or f for different types of surface are determined from experiments; see Streeter and Wylie (1975) for details.

In problems of surface flow which involves flooding and drying Falconer and Chen (1991) point out the advantages of using the Nikuradse parameter k_s to characterize the surface roughness. This is defined as the sand grain diameter for a sand-coated surface with the same f value. French (1987) gives a modified Moody diagram relating the value of f to the Reynolds' number:

$$\mathbf{R}_e = \frac{k_s}{\nu} \sqrt{(gRs)}$$

(where ν is the kinematic viscosity) for a range of values of $2R/k_s$. The usefulness of the parameter k_s is that it can be related directly to the minimum depth. If the computed average depth over an area is less than this, the corresponding numerical values of depth and velocity are equated to zero to represent the drying out of this area.

Steady-flow problems can often be represented by fairly simple formulae derived from the equations of continuity and momentum. These can then be solved directly by easy integration or trial and error methods as described in textbooks such as French (1987) or Streeter and Wylie (1975) and are not considered further here. The problems considered in this book lead to the types of flow equations which are too complicated to be solved directly and have to be treated by numerical methods.

1.3 Surface flow—lateral velocity distribution

This problem is chosen for a start because it is a boundary value problem leading to an ordinary differential equation (see Table 1.1.1). This particularly relates to the flow in a compound channel, i.e. a main channel plus flood plains on either side. The aim is to be able to predict the stage–discharge relationship (stage = height of water surface above datum). Estimates based on one-dimensional flow theory taking the channel as a single unit can give flows 30% too low; divided channel methods splitting into channel and flood plains give estimates which are too high (Wark *et al.*, 1990). The following method takes account of the non-uniformity of flow across the channel. It is assumed that the channel is straight, the flow is steady and uniform in the direction of the channel, and the water surface across the channel is horizontal. The required equation can be derived from the general two-dimensional shallow-water equations (French, 1987) or, more simply, from first principles as follows. Figure 1.3.1 shows a cross-section through the channel. The x axis is taken horizontally across the channel which

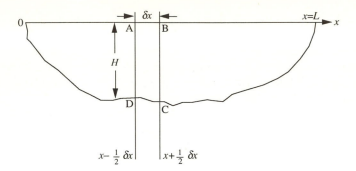

Figure 1.3.1: Cross-section of channel.

extends from $x = 0$ to $x = L$. ABCD is the cross-section of a slice of width δx, average depth H, and down-channel length unity. Then because the flow is steady there must be a balance between the forces of gravity, friction, and the momentum exchange due to turbulence acting on the slice of water. The turbulence effect is $\varepsilon H \, du/dx$ evaluated on the sides of the slice where ε is a turbulence exchange parameter and u the depth-averaged velocity. Balancing the forces gives

$$\rho g H \delta x s - F \delta x + \rho \left[\varepsilon H \frac{du}{dx} \right]_{x+\frac{1}{2}\delta x} - \rho \left[\varepsilon H \frac{du}{dx} \right]_{x-\frac{1}{2}\delta x} = 0$$

i.e.

$\rho g H \delta x s - F \delta x$

$$+ \rho \left[\varepsilon H \frac{du}{dx} + \tfrac{1}{2}\delta x \frac{d}{dx}\left(\varepsilon H \frac{du}{dx} \right) - \left\{ \varepsilon H \frac{du}{dx} - \tfrac{1}{2}\delta x \frac{d}{dx}\left(\varepsilon H \frac{du}{dx} \right) \right\} \right] = 0 \quad (1.3.1)$$

where ρ is the density, g the acceleration due to gravity, s the slope of the channel downstream, and F the friction force per unit area.

This leads to

$$\frac{d}{dx}\left(\varepsilon H \frac{du}{dx} \right) = \frac{k|u|u}{\rho} - gHs \quad (1.3.2)$$

taking $F = k|u|u$ where $|u|$ is the numerical value of u. (A useful check on the signs of the terms in (1.3.2) is to recall that u is expected to have a maximum value in the main channel; hence d^2u/dx^2 is negative and the gravity force must be greater than the friction.)

For a particular value of the stage, $H(x)$ is known; thus the corresponding $u(x)$ and hence the discharge can be calculated from the solution of eqn (1.3.2). Equation (1.3.2) is an ordinary differential equation representing a boundary value problem as in Table 1.1.1. The boundary values are $u = 0$ at the sides of the channel, $x = 0, L$. The solution of the kind of equation given by (1.3.2) by finite differences is discussed in section 4.1.

1.4 Shallow-water equations

This kind of surface flow is characterized as gradually varying unsteady flow. The angle of slope of the base of the flow is small, the change of depth with time is slow, the curvature of the wave profile is small, it is sufficient to take an averaged velocity, and the effect of boundary friction is included. The methods described here apply to free surface flows and are commonly used for the representation of flood waves, tidal flows, and storm surges.

The typical equations of surface flow can be illustrated by considering the shallow-water flow which is the effect of rainfall on a gentle slope with an impermeable surface. The flow considered here is what is called 'sheet flow' where the depth of water is small compared with the breadth of the slope over which it is flowing. Consider an element of fluid of constant breadth, b, depth H, and length dx which is at a distance x from the top of the slope as shown in Fig. 1.4.1. If the mean velocity in this element is u, the discharge is $q = uHb$, and if r represents the rainfall (in millimetres per hour, for example) the balance, net increase of water in the element in time dt equals water in minus water out in time dt, gives the continuity equation

$$\left\{\left(H + \frac{1}{2}\frac{\partial H}{\partial t}\,dt\right) - \left(H - \frac{1}{2}\frac{\partial H}{\partial t}\,dt\right)\right\}b\,dx$$

$$= \left\{\left(q - \frac{1}{2}\frac{\partial q}{\partial x}\,dx\right) - \left(q + \frac{1}{2}\frac{\partial q}{\partial x}\,dx\right)\right\}dt + br\,dx\,dt$$

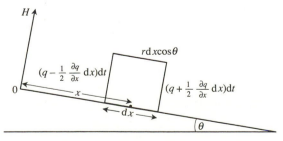

Figure 1.4.1: Flow down slope—continuity.

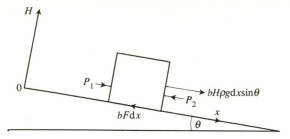

Figure 1.4.2: Flow down slope—balance of forces.

(assuming the slope is small enough to put $\cos \theta \approx 1$), i.e.

$$b\frac{\partial H}{\partial t} + \frac{\partial q}{\partial x} = br. \tag{1.4.1}$$

Putting $q = uHb$, this equation is equivalent to

$$\frac{\partial H}{\partial t} + H\frac{\partial u}{\partial x} + u\frac{\partial H}{\partial x} = r. \tag{1.4.2}$$

There is also the momentum equation obtained by balancing the forces on the element. From Fig. 1.4.2, assuming hydrostatic pressures we have: force acting on a cross-section of breadth b is

$$b\int_0^H \rho gy\ dy = \tfrac{1}{2}b\rho gH^2$$

where ρ is the density and g is the acceleration due to gravity. Hence the pressure forces on the upstream and downstream faces of the element in Fig. 1.4.2 are

$$P_1 = \tfrac{1}{2}\rho gb\left(H - \frac{1}{2}\frac{\partial H}{\partial x}dx\right)^2, \qquad P_2 = \tfrac{1}{2}\rho gb\left(H + \frac{1}{2}\frac{\partial H}{\partial x}dx\right)^2.$$

Thus the equation of motion of the element is given by equating the rate of change of momentum to the difference in the pressure forces plus gravity minus the friction, i.e.

$$\frac{d}{dt}(Mu) = M\frac{du}{dt} + u\frac{dM}{dt} = P_1 - P_2 + \rho Hbg\ dxs - Fb\ dx$$

where M is the mass of the element, $s = \sin \theta$, and F is the friction force per unit area. Substituting $M = \rho Hb\ dx$, using

$$\frac{du}{dt} = \frac{\partial u}{\partial t} + u\frac{\partial u}{\partial x} \quad \text{and} \quad \frac{dM}{dt} = \rho rb\ dx$$

we have

$$M\frac{du}{dt} + u\frac{dM}{dt} = \rho Hb\,dx\left(\frac{\partial u}{\partial t} + u\frac{\partial u}{\partial x}\right) + u\rho rb\,dx.$$

Then substituting for P_1 and P_2 and cancelling $\rho b\,dx$ gives the momentum equation:

$$uH\frac{\partial u}{\partial x} + H\frac{\partial u}{\partial t} + gH\frac{\partial H}{\partial x} + ru - gHs + \frac{F}{\rho} = 0 \qquad (1.4.3)$$

or

$$uH\frac{\partial u}{\partial x} + H\frac{\partial u}{\partial t} + gH\frac{\partial h}{\partial x} + ru + \frac{F}{\rho} = 0$$

where h is the height of the water surface. Introducing the 'friction' slope $s_f = F/(g\rho H)$ (section 1.2), the momentum equation can be written as

$$s_f = s - \left(\frac{\partial H}{\partial x} + \frac{u}{g}\frac{\partial u}{\partial x} + \frac{1}{g}\frac{\partial u}{\partial t} + \frac{ru}{gH}\right). \qquad (1.4.4)$$

Similarly the continuity equation for the vertically integrated two dimensions in plan flow assuming no influx is

$$\frac{\partial q_x}{\partial x} + \frac{\partial q_y}{\partial y} + \frac{\partial h}{\partial t} = 0 \qquad (1.4.5)$$

where the x and y axes are respectively taken along and across the channel, q_x and q_y are the components of volume flux per unit time and unit width of the vertical cross-sections perpendicular to the x and y directions respectively, and h is the height of the water surface above some datum. If u, v are the depth-averaged components of the velocity in the x and y directions respectively, and H is the depth of the flow, then putting $q_x = Hu$, $q_y = Hv$, an alternative form of eqn (1.4.5) is

$$\frac{\partial}{\partial x}(Hu) + \frac{\partial}{\partial y}(Hv) + \frac{\partial h}{\partial t} = 0. \qquad (1.4.6)$$

(Equation (1.4.5) can be obtained by integrating the three-dimensional continuity equation through the vertical using Leibniz's rule to incorporate the boundary conditions at top and bottom as for the groundwater flow equations in Appendix 1.)

The corresponding momentum equations are

$$\frac{\partial}{\partial t}(uH) + \left[\frac{\partial}{\partial x}(u^2H) + \frac{\partial}{\partial y}(uvH)\right] + gH\frac{\partial h}{\partial x} + \text{friction terms} = 0 \quad (1.4.7a)$$

and

$$\frac{\partial}{\partial t}(vH) + \left[\frac{\partial}{\partial x}(uvH) + \frac{\partial}{\partial y}(v^2H)\right] + gH\frac{\partial h}{\partial y} + \text{friction terms} = 0. \quad (1.4.7b)$$

The momentum equations used by Falconer and Chen (1991) are of the form

$$\frac{\partial}{\partial t}(uH) + \beta\left[\frac{\partial}{\partial x}(u^2H) + \frac{\partial}{\partial y}(uvH)\right] + gH\frac{\partial h}{\partial x} - fvH + \frac{1}{\rho}(\tau_{bx} - \tau_{sx})$$

$$- \varepsilon H\left[\frac{\partial^2 u}{\partial x^2} + \frac{\partial^2 u}{\partial y^2}\right] = 0 \quad (1.4.8a)$$

and

$$\frac{\partial}{\partial t}(vH) + \beta\left[\frac{\partial}{\partial x}(uvH) + \frac{\partial}{\partial y}(v^2H)\right] + gH\frac{\partial h}{\partial y} + fuH + \frac{1}{\rho}(\tau_{by} - \tau_{sy})$$

$$- \varepsilon H\left[\frac{\partial^2 v}{\partial x^2} + \frac{\partial^2 v}{\partial y^2}\right] = 0 \quad (1.4.8b)$$

where f is the Coriolis parameter, ρ is the fluid density, τ_{bx}, τ_{by} are the bed shear stress components, τ_{sx}, τ_{sy} are the wind shear stress components on the surface, β is the correction factor to allow for a non-uniform vertical velocity profile, and ε is the depth-averaged turbulent eddy viscosity (see the ASCE Task Committee (1988) report for a full discussion of turbulence modelling). The introduction of the second-derivative terms in eqns (1.4.8) changes the equations to parabolic type so that information on u and v is now required all round the boundary of the region being studied.

If it is reasonable to assume no variation in the y direction across the channel, q_x is replaced by q/b and with the allowance for an influx r eqns (1.4.7) revert to the one-dimensional forms of the continuity and momentum equations (1.4.2) and (1.4.4). These are the St Venant equations. An alternative pair of equations suitable for channel flow can be obtained by substituting $u = q/A$ and $H = A/b$ where A is the wetted cross-sectional area, and $H = h - s(L - x)$ where L is the length of the slope (Fig. 1.4.1). Equation (1.4.1) becomes

$$\frac{\partial A}{\partial t} + \frac{\partial q}{\partial x} = br. \quad (1.4.9)$$

Substituting and multiplying through by b in the momentum equation (1.4.4) gives

$$A\frac{\partial}{\partial t}\left(\frac{q}{A}\right) + q\frac{\partial}{\partial x}\left(\frac{q}{A}\right) + gA\left(\frac{\partial h}{\partial x} + s_f\right) + br\frac{q}{A} = 0. \quad (1.4.10)$$

Expanding the first term on the left gives

$$\frac{\partial q}{\partial t} - \frac{q}{A}\frac{\partial A}{\partial t} + q\frac{\partial}{\partial x}\left(\frac{q}{A}\right) + gA\left(\frac{\partial h}{\partial x} + s_f\right) + br\frac{q}{A} = 0.$$

Rearranging and substituting for br from eqn (1.4.9) then gives

$$\frac{\partial q}{\partial t} + \frac{\partial}{\partial x}\left(\frac{q^2}{A}\right) + gA\left(\frac{\partial h}{\partial x} + s_f\right) = 0. \qquad (1.4.11)$$

The dimensionless friction term here is the friction slope as in section 1.2. For shallow sheet flow of depth H the cross-section area per unit width is taken as H and the 'wetted perimeter' is unity (supposing the depth H is negligible compared with the unit width).

The numerical solutions of these shallow-water equations are discussed in Chapters 2, 3, and 4. In this chapter we want to make further assumptions so that we can take a single kinematic wave equation to represent the sheet flow problem. This enables us to discuss the type of solution to be expected here by means of a simple example.

1.5 The kinematic wave equation

The single kinematic wave model of the sheet flow assumes that the inertia and pressure terms in eqn. (1.4.4) are negligible in comparison with the gravity and friction terms so that we can take

$$s_f = s. \qquad (1.5.1)$$

Then substituting into the general formula (1.2.4) gives

$$s = Cv^n/R^m \qquad (1.5.2)$$

which can be turned round to give a formula for the mean velocity v, assumed to be a single-valued function of R, as

$$v = \left[\frac{s}{C}R^m\right]^{1/n}. \qquad (1.5.3)$$

In terms of the discharge q and the depth H for sheet flow over a unit width, putting $R = A = H$ and $v = q/A = q/H$, gives

$$q = \left(\frac{s}{C}\right)^{1/n} H^{1+m/n}. \qquad (1.5.4)$$

Thus the Chezy formula with $n = 2$, $m = 1$ gives

$$q = \left(\frac{s}{C}\right)^{1/2} H^{3/2} \qquad (1.5.5)$$

and Manning with $n = 2$, $m = 4/3$ gives

$$q = \left(\frac{s}{C}\right)^{1/2} H^{5/3}. \tag{1.5.6}$$

In general these formulae produce a relationship of the form

$$q = \alpha H^{\beta} \tag{1.5.7}$$

where $\beta = 5/3$ for Manning or $\beta = 3/2$ for Chezy. We then have

$$\frac{\partial q}{\partial x} = \frac{\mathrm{d}q}{\mathrm{d}H} \frac{\partial H}{\partial x} = \beta \alpha H^{\beta-1} \frac{\partial H}{\partial x}. \tag{1.5.8}$$

Substituting from eqn (1.5.8) into eqn (1.4.1) gives

$$\frac{\partial H}{\partial t} + \beta \alpha H^{\beta-1} \frac{\partial H}{\partial x} = r \tag{1.5.9}$$

which is a form of the kinematic wave equation

$$\frac{\partial H}{\partial t} + c \frac{\partial H}{\partial x} = r \tag{1.5.10}$$

where $c = \mathrm{d}q/\mathrm{d}H$ is the kinematic wave speed.

Alternatively (1.5.1) can be taken to mean that the discharge q can be expressed as a single-valued function of the cross-sectional area A so that

$$\frac{\partial q}{\partial t} = \frac{\mathrm{d}q}{\mathrm{d}A} \frac{\partial A}{\partial t}. $$

Then substituting in the channel flow equation (1.4.5) with $b = 1$ gives another form of kinematic wave equation:

$$\frac{\partial q}{\partial t} + \hat{c} \frac{\partial q}{\partial x} = \hat{c}r $$

where $\hat{c} = \mathrm{d}q/\mathrm{d}A$, which is of the same basic pattern as eqn (1.5.10).

First consider the solution of eqn (1.5.10) with a constant value of c and a constant rate of rainfall r in order to demonstrate the form of the solution of this type of equation. Substitute

$$\frac{\partial H}{\partial x} = P \quad \text{and} \quad \frac{\partial H}{\partial t} = Q. \tag{1.5.11}$$

Then eqn (1.5.10) becomes

$$Q + cP = r. \tag{1.5.12}$$

Now because the depth of the water H is a function of x and t we have the incremental equation

$$\mathrm{d}H = \frac{\partial H}{\partial x}\,\mathrm{d}x + \frac{\partial H}{\partial t}\,\mathrm{d}t = P\,\mathrm{d}x + Q\,\mathrm{d}t. \tag{1.5.13}$$

Eliminating P between eqns (1.5.12) and (1.5.13) gives

$$Q(c\,\mathrm{d}t - \mathrm{d}x) + r\,\mathrm{d}x - c\,\mathrm{d}H = 0. \tag{1.5.14}$$

On curves where the bracket on the left-hand side of eqn (1.5.14) vanishes we have

$$\frac{\mathrm{d}x}{\mathrm{d}t} = c, \quad \text{i.e. } x = ct + \text{constant.}$$

These curves are called characteristics. The characteristics are the key to the kind of solution to be expected from this type of equation because they 'carry' the information about the solution. On the characteristics the equation

$$x - ct = \text{constant}$$

is satisfied, together with (from eqn. (1.5.14)) the simple relationship

$$\frac{\mathrm{d}H}{\mathrm{d}x} = \frac{r}{c}. \tag{1.5.15}$$

The equation of the characteristic which goes through $x = x_1$ (less than L, where L is the length of the slope), when $t = 0$, is

$$x = ct + x_1 \quad \text{or} \quad t = (x - x_1)/c \tag{1.5.16}$$

On this characteristic, using eqn (1.5.15), the solution of eqn (1.5.10) is

$$H = \frac{r}{c}(x - x_1) = rt \tag{1.5.17}$$

supposing that the rainfall starts at $t = 0$ and before that there was no water on the slope. On the characteristic which goes through $x = 0$, $t = 0$ we have

$$x = ct \quad \text{and} \quad H = rt = rx/c.$$

On the characteristic which goes through $x = 0$, $t = t_1$ we have

$$x = c(t - t_1) \quad \text{and} \quad H = rx/c = r(t - t_1) \tag{1.5.18}$$

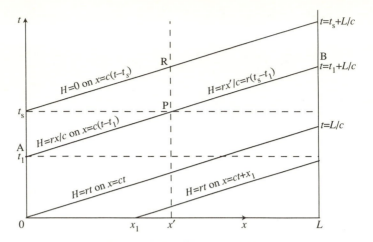

Figure 1.5.1: Characteristics.

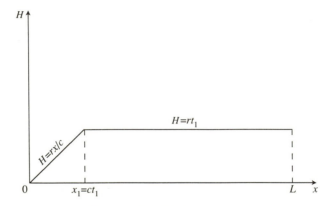

Figure 1.5.2: Profile at $t = t_1$.

up to the time $t = t_s$ when the rain stops. Figure 1.5.1 shows the characteristics for this problem.

We can get the profile of the water surface at time $t = t_1$ by taking the H values at the points with $t = t_1$ as shown in Fig. 1.5.2 for $t_1 < t_s$. Figure 1.5.3 shows the values of H corresponding to a particular value of $x = x'$ which gives the variation in time of the depth of water at this point on the slope. The discharge at the bottom of the slope where $x = L$ is $q(L) = cH(L)$ and from Fig. 1.5.2 this becomes steady at $q = rL$ when $t = L/c$ provided $t_s > L/c$. Then the rate of discharge from the bottom of the slope equals the rate at which the rain is falling on the slope.

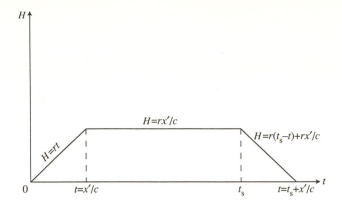

Figure 1.5.3: Variation of depth with time as seen at $x = x'$.

Looking in detail at Fig. 1.5.1, on characteristic APB, the part AP is conveying the information that it was still raining at $x = 0$ at $t = t_1$, and hence $H = rx'/c$ at $x = x'$, but at P the rain stops and for $t > t_s$ we are solving $dH/dx = 0$, i.e. $H = $ constant and hence $H = rx'/c$ on PB. At R, where the $x = x'$ line meets the characteristic through $t = t_s$, we have the depth $H = 0$. Figure 1.5.3 illustrates the effect often graphically described as a 'wall' of water advancing down the slope in serious flooding. The solution of the kinematic wave equation (1.5.10) has been simplified here by assuming $\beta = 1$ in eqn (1.5.7) and constant rainfall, but the solution of eqn (1.5.10) is similar with curves instead of the straight lines as shown in Fig. 1.5.4. This shows the discharge at the bottom of the slope due to steady rainfall of 10 mm per hour for 0.3 hours using the Chezy formula for the friction and is obtained by methods described in Chapter 2.

Chapter 3 includes the finite difference schemes for the solution of the kinematic wave equation and Chapter 4 discusses the solution of the full St Venant equations for sheet and channel flow. Chapter 7 includes cases where the surface flow is combined with subsurface flow and we have the more complicated situation where some of the rain goes into the ground and may emerge farther down the slope because the ground is saturated there.

Any equation which has characteristic curves is called hyperbolic. The example above demonstrates how in a sense the characteristics can be said to carry the information from one point of (x, t) space to another.

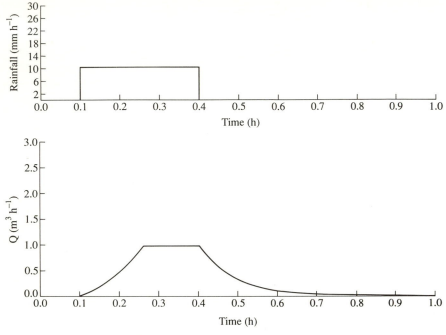

Figure 1.5.4: Discharge at bottom of slope.

1.6 Pollution transport in rivers

Another example of the hyperbolic type of equation is given by the continuity equation for the transport of a pollutant of concentration C in a river flowing with constant average velocity c. This is obtained in the same way as the continuity equation in section 1.4. from: net increase of pollutant in an infinitesimal element of fluid in time dt equals pollutant in minus pollutant out in time dt.

The result is of the same form as eqn (1.5.10), i.e.

$$\frac{\partial C}{\partial t} + c \frac{\partial C}{\partial x} = 0 \tag{1.6.1}$$

where x is the distance along the river. Assume the pollutant is of the same density as the water. Suppose a factory at $x = 0$ on the river bank discharges a slug of effluent at time $t = 0$ and that there is instantaneous mixing across the cross-section of the river so that the problem is one dimensional. Of course this mixing cannot actually be instantaneous but the assumption is a convenient approximation. Also assume initially that no diffusion of the effluent takes place. Then the effluent

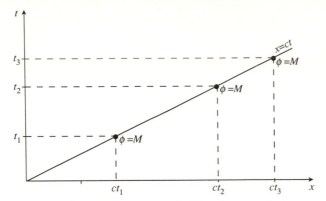

Figure 1.6.1: Effluent carried by characteristics.

is 'carried' by the characteristic as shown in Fig. 1.6.1 and the observer at a distance L downstream will see the slug of effluent passing at time L/c. This illustration is useful when we are discussing the numerical solution of this kind of equation.

When the effluent does diffuse into the water in the river, the equation of effluent transport becomes

$$\frac{\partial C}{\partial t} = D\frac{\partial^2 C}{\partial x^2} - c\frac{\partial C}{\partial x} \qquad (1.6.2)$$

where D is the coefficient of diffusion. This equation is now of a type called parabolic. If it is assumed that the slug of effluent is discharged into still water where $c = 0$, the equation is the one-dimensional diffusion equation

$$\frac{\partial C}{\partial t} = D\frac{\partial^2 C}{\partial x^2}. \qquad (1.6.3)$$

Supposing an amount V of effluent is discharged at $x = 0$ at $t = 0$, the solution of eqn (1.6.3) is given by

$$C = \frac{V}{2\sqrt{(\pi Dt)}}\exp\left(-\frac{x^2}{4Dt}\right) \qquad (1.6.4)$$

(Crank, 1975). If $x \neq 0$, then when $t = 0$, $C = 0$ (because the exponential in (1.6.4) dominates in the limit). When $x = 0$ and $t = 0$, C is infinite but in such a way that when we integrate over the whole region to obtain the total amount of effluent present at any time including $t = 0$, this is equal to V as expected. This device of integrating out a 'singularity' is also used to deal with abstraction and recharge wells in groundwater flow as in Chapters 4 and 5. This can be demonstrated as follows.

Substituting from eqn (1.6.4), and putting $x = 2\sqrt{(Dt)}\eta$, and hence $dx = 2\sqrt{(Dt)}\,d\eta$, we have

$$\int_{-\infty}^{\infty} C(x, t)\,dx = \frac{V}{\sqrt{\pi}} \int_{-\infty}^{\infty} \exp(-\eta^2)\,d\eta. \tag{1.6.5}$$

Now the error function is given by

$$\operatorname{erf}(z) = \frac{2}{\sqrt{\pi}} \int_{0}^{z} \exp(-\eta^2)\,d\eta \tag{1.6.6}$$

and we know (from e.g. Crank, 1975) that $\operatorname{erf}(\infty) = 1$. Hence

$$\frac{1}{\sqrt{\pi}} \int_{-\infty}^{\infty} \exp(-\eta^2)\,d\eta = \frac{2}{\sqrt{\pi}} \int_{0}^{\infty} \exp(-\eta^2)\,d\eta = 1$$

and from eqn (1.6.5) we have therefore

$$\int_{-\infty}^{\infty} C(x, t)\,dx = V$$

for all values of t. Figure 1.6.2 shows the distribution of C/V against x for $Dt = 0.0625$, 0.5, 1.0. The expression (1.6.4) shows that the discharge of effluent has an immediate effect throughout the length of water, though, of course, this is very small at large distances. This is typical of parabolic equations and quite different from the solution of the hyperbolic equation for the convective transport (1.5.1) where the observer on the river bank has to wait for the effluent to arrive 'on the characteristic'.

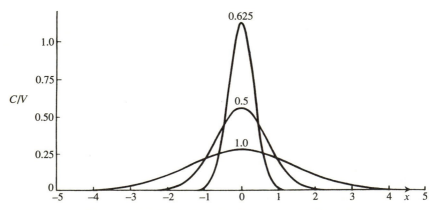

Figure 1.6.2: Diffusion of effluent into still water.

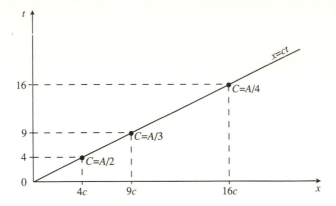

Figure 1.6.3: Advection–diffusion of effluent.

For the solution of the combined convection and diffusion equation (1.6.2) we make the substitution $x' = x - ct$ to give

$$\frac{\partial C}{\partial t} = D \frac{\partial^2 C}{\partial x'^2} \qquad (1.6.7)$$

so that the solution of (1.6.2) can now be seen to be

$$C = \frac{V}{2\sqrt{(\pi Dt)}} \exp\left(-\frac{(x - ct)^2}{4Dt}\right). \qquad (1.6.8)$$

On the line $x = ct$ we can write $C = At^{-0.5}$, where A is independent of t. Figure 1.6.3 shows the decrease of the concentration of the effluent along the $x = ct$ line as it diffuses up- and downstream during its transport by the current.

Some problems with hyperbolic equations such as (1.5.10) and (1.6.1) can be solved by the method of characteristics described in Chapter 2. The most popular methods for the solution of the surface flow equations are the finite difference methods described in Chapter 3.

1.7 Groundwater flow

For the details of the flow in porous media the reader is referred to textbooks such as Lencastre (1987) for the engineering approach and to Bear (1979) for a very comprehensive and detailed survey. In this section we shall just summarize the theory in order to form the typical flow equations and discuss the type of solution to be expected.

In a porous medium the piezometric head is defined as

$$\phi = \text{sum of elevation head plus pressure head.}$$

For an incompressible fluid this is

$$\phi = z + p/\gamma$$

where z is the height above datum, γ is the specific weight of the fluid, and p is the pressure. The water moves because of the difference in piezometric head between one point and another. The specific discharge, $\mathbf{q}$, is defined as the volume of water flowing per unit time through a unit cross-sectional area normal to the direction of the flow. This represents an averaged velocity of the filtration of the water through the interstices of the medium and is related by Darcy's law (Darcy, 1856) to the gradient of the piezometric potential or head of water ϕ by the equation

$$\mathbf{q} = -K \operatorname{grad} \phi = -K\nabla\phi \qquad (1.7.1)$$

(throughout this book the shorthand notation on the right of (1.7.1) is used for the gradient of a scalar function—see the end of the section on notation for the extended form). In (1.7.1) K is a property of the permeability of the porous medium known as the hydraulic conductivity and has the dimension LT^{-1}. The velocities in groundwater flow are in general very low; the Reynold's number, $\mathbf{R_e} = qd/v$, where q is the magnitude of the specific discharge, d is a representative length in the porous medium through which the water flows, and v is the kinematic viscosity of the water, has generally some value between 1 and 10 for Darcy's law to be valid (Bear, 1979).

The equation of continuity is obtained by using the Eulerian approach of calculating the mass balance in an element centred at a point (x, y, z) as shown in Fig. 1.7.1, supposing that the mass flux vector is $\mathbf{J} = (J_1, J_2, J_3)^T$. Then mass in minus mass out during time δt in the x direction is

$$\left\{ J_1\left(x - \frac{\partial x}{2}, y, z\right) - J_1\left(x + \frac{\partial x}{2}, y, z\right) \right\} \partial y\, \partial z$$

$$= \frac{\partial J_1}{\partial x} \partial x\, \partial y\, \partial z + \text{higher-order terms.}$$

Hence, taking all three directions, mass in minus mass out per unit volume per unit time as $\partial x, \partial y, \partial z \to 0$ is

$$-\left(\frac{\partial J_1}{\partial x} + \frac{\partial J_2}{\partial y} + \frac{\partial J_3}{\partial z} \right) = -\operatorname{div} \mathbf{J} \equiv -\nabla\cdot\mathbf{J}.$$

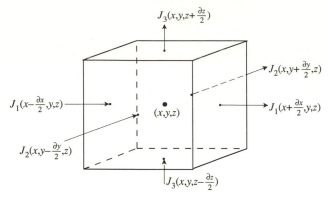

Figure 1.7.1: Equation of continuity for subsurface flow.

(Throughout this book the shorthand notation on the right is used for the divergence of a vector—see the end of the section on notation for the extended form.) Now $\mathbf{J} = \rho\mathbf{q}$ and the change in mass per unit volume per unit time is $(\partial/\partial t)(\rho n)$, where n is the porosity of the ground. This is replaced by $\rho S_0\, \partial\phi/\partial t$ where ϕ is the piezometric head and S_0 is the specific storativity defined as the volume of water released or stored per unit volume per unit change in the piezometric head.

Bear (1979) discusses the details of this replacement and observes that in practice they do not matter as the value of S_0 is expected to be obtained from site measurements. Thus the equation of continuity for groundwater flow is given by

$$S_0\frac{\partial\phi}{\partial t} = -\mathbf{\nabla}\cdot\mathbf{q}. \tag{1.7.2}$$

From eqns (1.7.1) and (1.7.2) we obtain the basic equation of time-dependent groundwater flow which is the parabolic equation:

$$S_0\frac{\partial\phi}{\partial t} = \mathbf{\nabla}(K\mathbf{\nabla}\phi). \tag{1.7.3}$$

This general equation (1.7.3) involves time and three space dimensions; it can be simplified in various ways:

1. In general the term K in (1.7.3) is a tensor. It is necessary for the study of flow in fissures in the ground which may be in any direction to use the general form

$$\mathbf{\nabla}K\mathbf{\nabla}(\phi) \equiv \frac{\partial}{\partial x_i}\left(K_{ij}\frac{\partial\phi}{\partial x_j}\right) \tag{1.7.4}$$

with coordinates x_1, x_2, x_3 and the tensor summation convention on repeated subscripts. But it is frequently considered sufficient to assume that the principal components of hydraulic conductivity are in the directions of the axes and can be taken as (K_x, K_y, K_z).

2. It is often considered sufficient to solve the steady-state elliptic boundary value problem (Table 1.1.1)

$$\mathbf{V}(K\mathbf{V}\phi) = 0. \tag{1.7.5}$$

(a) If it is a problem where it is possible to assume no variation in one horizontal direction (1.7.5) reduces to the two-dimensional elliptic equation

$$\frac{\partial}{\partial x}\left(K_x \frac{\partial\phi}{\partial x}\right) + \frac{\partial}{\partial z}\left(K_z \frac{\partial\phi}{\partial z}\right) = 0. \tag{1.7.6}$$

(i) If the ground is assumed saturated up to a certain level with no water above, this level is called the phreatic surface or water table. Two boundary conditions are needed to fix the position of this:
(1) zero normal flow, i.e.

$$K_x \frac{\partial\phi}{\partial x} n_1 + K_z \frac{\partial\phi}{\partial z} n_2 = 0 \tag{1.7.7a}$$

where (n_1, n_2) is the unit outward normal to the phreatic surface;
(2) the pressure is atmospheric (taken as $p = 0$), i.e.

$$\phi = z. \tag{1.7.7b}$$

Equation (1.7.6) is frequently further simplified by assuming the ground is isotropic, i.e. $K_x = K_z = K$.

(ii) If water is assumed present between the saturated layer and the ground surface the equation used is Richards' equation (Richards, 1931) for saturated–unsaturated flow with empirical relations for K and for S_0. This is discussed in Chapter 7. In general there is a hysteresis effect, i.e. the variations of K and S_0 in the unsaturated zone depend on whether the medium is undergoing wetting or drying. This effect is usually neglected.

3. As shown in Appendix 1, eqn (1.7.3) can be integrated through the vertical, z, dimension to give the equation

$$S\frac{\partial h}{\partial t} = \frac{\partial}{\partial x}\left(\bar{K}_x(h - h_b)\frac{\partial h}{\partial x}\right) + \frac{\partial}{\partial y}\left(\bar{K}_y(h - h_b)\frac{\partial h}{\partial y}\right) \tag{1.7.8a}$$

where h, h_b are the heights of the water table and the base of the aquifer respectively and $\bar{K}_x$, $\bar{K}_y$ are the vertically averaged components of the hydraulic conductivity in the x and y directions respectively. The boundary conditions on the top and bottom of the aquifer have been incorporated into eqns (1.7.8a) by using Leibniz's rule. A further simplification is to introduce the transmissivity components:

$$T_x = \int_{h_b}^{h} K_x \, dz, \qquad T_y = \int_{h_b}^{h} K_y \, dz \qquad (1.7.8b)$$

which can be estimated from site pumping tests (Fig. 1.7.2). The result is the equation

$$S \frac{\partial h}{\partial t} = \frac{\partial}{\partial x}\left(T_x \frac{\partial h}{\partial x}\right) + \frac{\partial}{\partial y}\left(T_y \frac{\partial h}{\partial y}\right). \qquad (1.7.9)$$

If a fully saturated aquifer with an impermeable layer top and bottom (a confined aquifer) is assumed to be homogeneous and isotropic

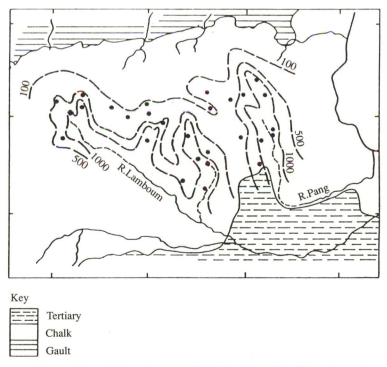

Key

≡≡≡	Tertiary
⬜	Chalk
☰	Gault

Figure 1.7.2: Areal variation of transmissivity determined from pumping tests.

and of constant depth B, eqn (1.7.3) can be integrated vertically to obtain

$$S\frac{\partial\phi}{\partial t} = T\left(\frac{\partial^2\phi}{\partial x^2} + \frac{\partial^2\phi}{\partial y^2}\right) \tag{1.7.10}$$

using now the effective storativity S and the transmissivity $T = KB$. Note that eqns (1.7.9) and (1.7.10) are of the same parabolic type with the head of water in the confined aquifer in one and the height of the water table in the other.

The steady-state forms of (1.7.9) and (1.7.10) are elliptic boundary value problems (Table 1.1.1).

When abstraction or recharge wells are present in the aquifer these can be represented in the numerical solution as described in Chapters 5, 7, and 8. It can also be useful to know analytic solutions for flow round a well and these are described in section 1.8.

This section concludes with some general remarks about aquifer models.

1. Choice of boundaries: suppose there is a region of an aquifer of particular interest which it is required to model. In practice this is rarely conveniently surrounded by boundaries on which the required boundary conditions are available (these must be known head or known flow). It is often necessary to model a larger area in order to have suitable boundaries. Section 7.3 gives examples of this.

2. Parabolic equations such as (1.7.3), (1.7.9), and (1.7.10) must also have initial conditions specified. These can frequently be obtained from a corresponding steady-state solution. But some problems such as that of flow in a hillslope described in section 7.9 require to be run-in in a controlled way before the simulation of interest starts.

1.8 Effect of an abstraction well

The theory in this section is included because it is useful in conjunction with the numerical methods described in later chapters. The steady flow to an abstraction well pumping at a constant rate in a phreatic aquifer can be approximated simply by using the Dupuit assumptions (Dupuit, 1863) of essentially horizontal flow (Appendix 1). With the bottom of the aquifer assumed horizontal and taken as datum, i.e. $h_b = 0$, the steady-state form of eqn. (A1.18) is

$$\nabla^2(\bar{K}(h^2/2)) = 0 \tag{1.8.1}$$

where h is the height of the phreatic surface (water table) and $\bar{K}$ is the vertically averaged hydraulic conductivity. In polar coordinates for the radial flow round a well in the centre of the region, h^2 is a function of the radial distance r only and the equation of continuity becomes

$$\frac{1}{r}\frac{d}{dr}\left[\bar{K}r\frac{d}{dr}\left(\frac{h^2}{2}\right)\right] = 0, \quad \text{for } r > r_w \tag{1.8.2}$$

where r_w is the radius of the well. The general solution of eqn (1.8.2) with constant $\bar{K}$ is of the form

$$h^2 = A\ln(r/r_w) + B \tag{1.8.3}$$

where A and B are to be determined from the boundary conditions. If there is a constant abstraction rate Q_w this must equal the flow across the perimeter of any circle of radius $r > r_w$, hence

$$Q_w = 2\pi r\bar{K}\frac{d}{dr}\left(\frac{h^2}{2}\right). \tag{1.8.4}$$

Substituting from (1.8.3) gives

$$A = Q_w/(\pi\bar{K}). \tag{1.8.5}$$

Then if $h = h_w$ at $r = r_w$ on the well surface (neglecting the seepage face effect)

$$h^2 - h_w^2 = \frac{Q_w}{\bar{K}\pi}\ln\left(\frac{r}{r_w}\right), \quad r > r_w \tag{1.8.6}$$

Also, if the height of the water table is known to be $h = h_0$ at a distance $r = R$, it follows that

$$h^2 - h_0^2 = \frac{Q_w}{\bar{K}\pi}\ln\left(\frac{r}{R}\right). \tag{1.8.7}$$

Combining eqns (1.8.6) and (1.8.7) gives

$$\frac{h_0^2 - h^2}{h_0^2 - h_w^2} = \frac{\ln(R/r)}{\ln(R/r_w)}. \tag{1.8.8}$$

These equations are only valid in the vicinity of the well. Figure 1.8.1 shows the approximate Dupuit cone of depression round a well. The distance R at which the drawdown is effectively zero is called the radius of influence; this parameter has to be estimated from site observations.

With a confined aquifer, supposing the bottom is level and the thickness $h - h_b$ is constant, and with constant hydraulic conductivity

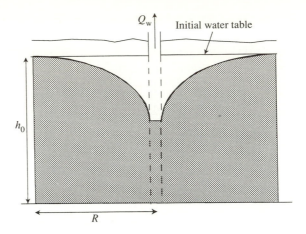

Figure 1.8.1: Approximate cone of depression.

so that the transmissivity $T = \bar{K}(h - h_{\mathrm{b}})$ is constant from (1.7.8b), the steady-state form of (1.7.10) gives

$$\nabla^2 \phi = 0 \qquad\qquad (1.8.9)$$

i.e. for axially symmetric flow round a well

$$\frac{\mathrm{d}}{\mathrm{d}r} r \frac{\mathrm{d}\phi}{\mathrm{d}r} = 0. \qquad\qquad (1.8.10)$$

Hence if $\phi = \phi_{\mathrm{w}}$ at the well radius $r = r_{\mathrm{w}}$ the solution corresponding to (1.8.6) is now

$$\phi = \phi_{\mathrm{w}} + \frac{Q_{\mathrm{w}}}{2\pi T} \ln\!\left(\frac{r}{r_{\mathrm{w}}}\right). \qquad\qquad (1.8.11)$$

Note that this simple solution only applies with the assumption of constant transmissivity.

The expression for the drawdown (1.8.3) is a function of r so that the start of the well pumping has theoretically an immediate effect throughout the region, but of course this is very small at large distances. This is typical of the parabolic equations in section 1.7 and we have here an effect similar to that with the diffusion equation (1.6.3). Eqn (1.8.2) has been formed on the assumption that the water is released from storage in the aquifer immediately upon a reduction in head. In practice there would be some delay, as explained by Bear (1979). A similar effect is to be expected from the solution of the equations representing groundwater flow in a hillslope. Rainfall entering the ground at the top of the hillslope will theoretically have an immediate effect on the discharge from the bottom of the hillslope.

Groundwater flow problems can be solved numerically by finite difference methods (Chapters 3 and 5), finite element methods (Chapters 6 and 7) and boundary integral methods (Chapter 8). Finite element methods are the most powerful and effective but they do require more effort to understand them and to use them to best advantage.

There are also problems of contaminant transport in groundwater. When the contaminant has the same density as the water we can solve the groundwater flow equation to obtain the potential ϕ and hence the velocity $\mathbf{v}$ to substitute into the transport equation. When the contaminant has a different density, salt for example, we have a more complicated problem with coupled equations. The numerical solution of these is discussed in Chapter 9.

2 The method of characteristics

2.1 Introduction

As explained in section 1.5 the hyperbolic equations such as the kinematic wave equation typically have curves associated with them called characteristics and these carry information about the solution. The method of characteristics uses this property to obtain the solution by moving out along the characteristics from the starting values. In some simpler cases it can be applied directly (section 2.2). In more complicated cases illustrated in sections 2.3 and 2.4 it is applied numerically.

In Chapter 1 the St Venant equations for surface flow are formulated. These can be in terms of the depth H and the mean velocity u as in eqns (1.4.2) and (1.4.3). Alternatively for channel flow they can be in terms of the cross-sectional area A and the discharge q as in (1.4.9) and (1.4.11). With further assumptions that the inertia and pressure terms are negligible compared with gravity and friction these two St Venant equations reduce to a single kinematic wave equation:

$$\frac{\partial H}{\partial t} + c\,\frac{\partial H}{\partial x} = r \qquad (2.1.1)$$

where c is the wave speed given by

$$c = \mathrm{d}q/\mathrm{d}H \qquad (2.1.2)$$

and r is the rainfall input per unit area.

Section 1.2 demonstrates the use of characteristics in the simple case where c is a constant and the characteristics are the straight lines

$$x = ct + x_1 \qquad (2.1.3)$$

when the characteristic starts from $x = x_1$, $t = 0$, or

$$x = c(t - t_1) \qquad (2.1.4)$$

when the characteristic starts from $x = 0$, $t = t_1$.

The method of characteristics described here can be used when c is no longer a constant but is derived from the assumption (1.5.1) that the friction slope and the channel slope are equal. A relationship of the form

$$q = \alpha H^m \qquad (2.1.5)$$

can then be used to derive c using (2.1.2):

$$c = c(H) = m\alpha H^{m-1}.$$ (2.1.6)

The kinematic wave equation is now

$$\frac{\partial H}{\partial t} + c(H)\frac{\partial H}{\partial x} = r.$$ (2.1.7)

The objective with the method of characteristics is to find a direction at each point in x, t space along which the integration of eqn (2.1.7) reduces to the integration of an ordinary differential equation only. In this direction the expression to be integrated will not depend on the partial derivatives in other directions.

Consider a general first-order hyperbolic equation of the form

$$a\frac{\partial H}{\partial x} + b\frac{\partial H}{\partial t} = f$$ (2.1.8)

where a, b, and f may be functions of H, x, and t.

Put $P = \partial H/\partial x$, $Q = \partial H/\partial t$; then eqn (2.1.8) becomes

$$aP + bQ = f.$$ (2.1.9)

Since H is a function of x and t,

$$dH = P\,dx + Q\,dt.$$ (2.1.10)

Eliminating P between eqns (2.1.9) and (2.1.10) gives

$$dH = \frac{(f - bQ)}{a}\,dx + Q\,dt$$ (2.1.11)

and rearranging produces

$$Q(a\,dt - b\,dx) + f\,dx - a\,dH = 0.$$ (2.1.12)

Along the curve where $a\,dt - b\,dx = 0$, i.e.

$$\frac{dx}{dt} = \frac{a}{b}$$ (2.1.13)

(a characteristic), eqn (2.1.12) reduces to

$$f\,dx = a\,dH.$$ (2.1.14)

The simplest way to write the conditions on a characteristic of eqn (2.1.8) is to combine eqns (2.1.13) and (2.1.14) in the form

$$\frac{dt}{b} = \frac{dx}{a} = \frac{dH}{f}.$$ (2.1.15)

On this characteristic the partial differential equation (2.1.8) reduces to an ordinary differential equation which may be taken as either

$$\frac{dH}{dt} = \frac{f}{b} \quad \text{or} \quad \frac{dH}{dx} = \frac{f}{a} \tag{2.1.16}$$

whichever is convenient.

2.2 Application of the method

The method of characteristics is now applied to the solution of eqn (2.1.8). Here

$$a = c(H), \qquad b = 1, \quad f = r \tag{2.2.1}$$

and r is assumed to represent a constant rate of rainfall. From (2.1.13) the characteristic curves are given by

$$\frac{dx}{dt} = c(H) = m\alpha H^{m-1} \tag{2.2.2}$$

It is supposed that the rain starts at $t = 0$ and prior to this there is a constant depth of water $H = H_0$ on the slope. Also it is assumed that for all time t, $H = H_0$ at the top of the slope, $x = 0$. Then the method of characteristics is applied as follows.

Along a characteristic starting from $x = 0$, $t = t_1$, use the first equation of (2.1.16) to give

$$\frac{dH}{dt} = r, \quad \text{i.e. } H - H_0 = r(t - t_1). \tag{2.2.3}$$

Along a characteristic which starts from $x = x_1$, $t = 0$, use the first equation of (2.1.16) to give

$$H - H_0 = rt \tag{2.2.4}$$

and the second equation of (2.1.16) to give

$$\frac{dH}{dx} = \frac{r}{c(H)} \tag{2.2.5}$$

Using (2.1.6), integrating this equation gives

$$r \int_{x_1}^{x} dx = \alpha \int_{H_0}^{H} mH^{m-1} \, dH$$

i.e.

$$r(x - x_1) = \alpha(H^m - H_0^m) \tag{2.2.6}$$

or

$$H = \left(\frac{r(x - x_1)}{\alpha} + H_0^m\right)^{1/m}.$$ (2.2.7)

From (2.1.13) the characteristic curves are given by

$$\frac{dx}{dt} = c(H).$$

Thus from (2.1.6) and (2.2.3), the characteristic through $x = 0$, $t = t_1$ is given by

$$\frac{dx}{dt} = m\alpha H^{m-1} = m\alpha[r(t - t_1) + H_0]^{m-1}.$$

Integrating this gives

$$x = \frac{\alpha}{r}\{[r(t - t_1) + H_0]^m - H_0^m\}.$$ (2.2.8)

From (2.1.6) and (2.2.7) the characteristic through $x = x_1$, $t = 0$ is given by

$$x - x_1 = \frac{\alpha}{r}[(rt + H_0)^m - H_0^m].$$ (2.2.9)

Thus the characteristics and the solution on them can be obtained from eqns (2.2.8), (2.2.9), and (2.2.7). Figure 2.2.1 illustrates this for $H_0 = 0$, $m = 3/2$ (Chezy). Figure 2.2.1(a) is a sketch of the characteristics, now curved, and 2.2.1(b) shows the variation of depth at a fixed distance $x = x_1$ down the slope for a constant rate of rainfall r.

Figure 1.5.4 shows the discharge at the bottom of the slope obtained from the solution by the method of characteristics of eqn (2.1.7) using the Chezy friction law. The rainfall here is 10 mm per hour for 0.3 hours as indicated.

2.3 The numerical method of characteristics

The numerical method of characteristics applied to the solution of eqn (2.1.7) entails the approximate solution of eqns (2.1.16) by using incremental ratios instead of derivatives, i.e.

$$\Delta t = \frac{\Delta x}{c(H)} = \frac{\Delta H}{r}.$$ (2.3.1)

A start can be made from a point P, $x = x_P$, on the $t = 0$ line as shown

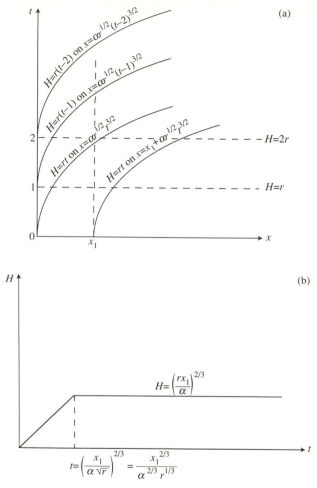

Figure 2.2.1: (a) Characteristics for example in section 2.2; (b) variation of depth at $x = x_1$.

in Fig. 2.3.1, and the aim is to approximate the characteristic PR and the solution along it.

Suppose R is the point $(x_P + \Delta x, \Delta t)$ where $H = H_0 + \Delta H$. Then in this case where there is just one family of characteristics, Δt is kept fixed and successive approximations $\Delta x^{(1)}, \Delta x^{(2)}, \ldots, \Delta H^{(1)}, \Delta H^{(2)}, \ldots,$ are found to Δx and ΔH. Equations (2.3.1) are used with $c(H)$ replaced by $c(H_0)$ as a first approximation:

$$\Delta x^{(1)} = c(H_0)\Delta t \quad \text{and} \quad \Delta H^{(1)} = r\Delta t.$$

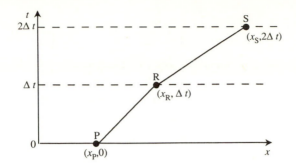

Figure 2.3.1: Numerical method of characteristics.

In succeeding iterations the non-linear term is replaced by averages:

$$\Delta x^{(2)} = \tfrac{1}{2}[c(H_0) + c(H_0 + \Delta H^{(1)})]\Delta t.$$

But with Δt fixed, $\Delta H^{(2)} = r\Delta t = \Delta H^{(1)}$; hence $\Delta x^{(3)} = \Delta x^{(2)}$ and the iteration is said to have converged.

Using (2.1.6) the converged values are

$$H_R = H_0 + r\Delta t \qquad (2.3.2)$$

and

$$x_R = x_P + (m\alpha/2)[H_0^{m-1} + (H_0 + r\Delta t)^{m-1}]\Delta t \qquad (2.3.3)$$

is the approximate equation of the characteristic.

Comparing with (2.2.4), eqn (2.3.2) is exact and subtracting (2.2.9) from (2.3.3) the error in the position of R is

$$x_R - x(\Delta t) = (m\alpha/2)[H_0^{m-1} + (H_0 + r\Delta t)^{m-1}]\Delta t - (\alpha/r)[(H_0 + r\Delta t)^m - H_0^m].$$

Supposing Δt can be chosen so that the dimensionless parameter $r\Delta t/H_0$ is small compared with unity (e.g. the effect of a rainfall which is small compared with the initial depth of water is being calculated), the brackets can be expanded to give the error in the position of R as

$$x_R - x(\Delta t) = \frac{m(m-1)(m-2)}{12}\,\alpha r^2 H_0^{(m-3)}\Delta t^3 + \text{higher powers of } \Delta t$$

$$= \varepsilon_1 + \text{higher powers of } \Delta t \qquad (2.3.4)$$

where ε_1 is the principal part of the error. This simple example illustrates the fact that the iteration is not necessarily going to converge to the exact value. With the Manning formula,

$$m = 5/3 \quad \text{and} \quad \varepsilon_1 = -\tfrac{5}{162}\alpha r^2 H_0^{-4/3}\Delta t^3.$$

With the Chezy formula,

$$m = 3/2 \quad \text{and} \quad \varepsilon_1 = -\tfrac{1}{32}\alpha r^2 H_0^{-3/2}\Delta t^3.$$

The value of Δt must be kept small to keep the error small.

The process can then be repeated in order to step out approximately along the characteristic from R to the point S where $t = 2\Delta t$. Suppose S is the point $(x_R + \Delta x, 2\Delta t)$; then eqn (2.3.1) gives

$$\Delta x^{(1)} = c(H_R)\Delta t = c(H_0 + r\Delta t) \quad \text{and} \quad \Delta H^{(1)} = r\Delta t.$$

Then

$$\Delta x^{(2)} = \tfrac{1}{2}[c(H_0 + r\Delta t) + c(H_0 + 2r\Delta t)]. \tag{2.3.5}$$

Again with Δt fixed, $\Delta H^{(2)} = \Delta H^{(1)} = r\Delta t$. Hence the iteration has converged and

$$x_S = x_R + \Delta x^{(2)}$$
$$= x_P + (m\alpha/2)[H_0^{m-1} + 2(H_0 + r\Delta t)^{m-1} + (H_0 + 2r\Delta t)^{m-1}]. \tag{2.3.6}$$

The error in the position of the point S calculated as before for the case where $r\Delta t/H_0$ is small compared with unity is obtained after some algebra:

$$x_S - x(2\Delta t) = 2\varepsilon_1 + \text{higher powers of } \Delta t. \tag{2.3.7}$$

Thus the error increases as the numerical method of characteristics proceeds.

If the parameter r here represents not a constant rainfall but a lateral inflow $r(x)$ to the channel which is a function of the distance x along the channel, then eqn (2.3.2) gives

$$\Delta x^{(1)} = c(H_0)\Delta t$$
$$\Delta H^{(1)} = \tfrac{1}{2}[r(x_1) + r(x_1 + \Delta x^{(1)})]\Delta t$$
$$\Delta x^{(2)} = \tfrac{1}{2}[c(H_0) + c(H_0 + \Delta H^{(1)})]\Delta t$$
$$\Delta H^{(2)} = \tfrac{1}{2}[r(x_1) + r(x_1 + \Delta x^{(2)})]\Delta t \ldots \text{etc.}$$

until convergence to a sufficient number of figures is reached. If the convergence is not fast enough, Δt should be decreased.

The problem just described has only the one family of characteristics shown in Fig. 2.2.1. The following is another example to illustrate the numerical method of characteristics applied to a pair of equations where there are now two families of characteristics.

Take a simplified dimensionless form of the St Venant equations:

$$\frac{\partial H}{\partial t} = \frac{\partial v}{\partial x}, \quad \frac{\partial v}{\partial t} = \frac{\partial H}{\partial x}. \tag{2.3.8}$$

Write

$$H_1 = \frac{\partial H}{\partial t} - \frac{\partial v}{\partial x} = 0, \quad H_2 = \frac{\partial v}{\partial t} - \frac{\partial H}{\partial x} = 0. \qquad (2.3.9)$$

Then it follows that

$$\lambda H_1 + H_2 = \lambda \left[\frac{\partial H}{\partial t} - \frac{1}{\lambda} \frac{\partial H}{\partial x} \right] + \left[\frac{\partial v}{\partial t} - \lambda \frac{\partial v}{\partial x} \right] = 0. \qquad (2.3.10)$$

Hence if $dx/dt = -1/\lambda$, the first square bracket in (2.3.10) equals

$$\frac{\partial H}{\partial t} + \frac{dx}{dt} \frac{\partial H}{\partial x} = \frac{dH}{dt}$$

and if $dx/dt = -\lambda$, the second square bracket in (2.3.10) equals

$$\frac{\partial v}{\partial t} + \frac{dx}{dt} \frac{\partial v}{\partial x} = \frac{dv}{dt}.$$

These two values of dx/dt are compatible if $\lambda^2 = 1$, i.e. $\lambda = \pm 1$. These two values of λ give the two families of characteristics. One of each of these characteristics is shown in Fig. 2.3.2. Along a characteristic eqn (2.3.10) reduces to

$$\lambda \frac{dH}{dt} + \frac{dv}{dt} = 0. \qquad (2.3.11)$$

That is, for $\lambda = -1$, $H - v = $ constant along the characteristics $x - t = a$ (constant), and for $\lambda = 1$, $H + v = $ constant along the characteristics $x + t = b$ (constant).

Take points P, Q at $x = 0.1, 0.2$ on the initial line $t = 0$ (Fig. 2.3.2). Suppose the initial values are given by $H = x^3, v = 0$. If the point R is the intersection of the characteristics $x - t = x_P$ through P and

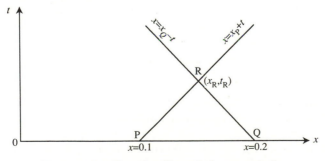

Figure 2.3.2: Two families of characteristics.

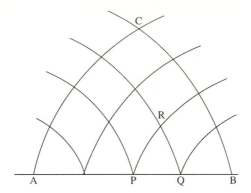

Figure 2.3.3: The domain of dependence.

$x + t = x_Q$ through Q, then the equations for the characteristics give

$$x_R - t_R = 0.1 \quad \text{and} \quad x_R + t_R = 0.2.$$

The solution is $x_R = 0.15$, $t_R = 0.05$. Also at R

$$H + v = 0.008, \quad H - v = 0.001.$$

Solving these equations gives

$$H_R = 0.0045, \quad v_R = 0.0035.$$

Of course as the equations (2.3.5) are linear this is the exact solution which is given by

$$H = \tfrac{1}{2}[(x + t)^3 + (x - t)^3], \quad v = \tfrac{1}{2}[(x + t)^3 - (x - t)^3].$$

Figure 2.3.3 shows the process continuing to calculate the solution at further points on the intersections of the two families of characteristics starting from points on the line AB and ending with the calculation of the value at C. The triangle ABC is called the domain of dependence for the solution at C because this solution depends on the values carried by the characteristics within this triangle.

The combination of two sets of characteristics together with iterating with average values of the non-linear terms is used in the more complicated problem of the numerical solution of the St Venant equations as shown in the next section.

2.4 The method of characteristics used for the numerical solution of the St Venant equations

The St Venant equations for subcritical flow in a channel supposing zero rainfall input can be written in the form

$$H_1 = \frac{\partial H}{\partial t} + H \frac{\partial v}{\partial x} + v \frac{\partial H}{\partial x} = 0 \qquad (2.4.1)$$

(from (1.4.2)) and

$$H_2 = \frac{\partial v}{\partial t} + v \frac{\partial v}{\partial x} + g \frac{\partial H}{\partial x} + g(s_f - s) = 0 \qquad (2.4.2)$$

(from 1.4.2)). Combining these two equations with a parameter λ then gives $\lambda H_1 + H_2 = 0$, where

$$\lambda H_1 + H_2 = (v + \lambda H) \frac{\partial v}{\partial x} + \frac{\partial v}{\partial t} + \lambda \left\{ \left(v + \frac{g}{\lambda} \right) \frac{\partial H}{\partial x} + \frac{\partial H}{\partial t} \right\} + g(s_f - s).$$
$$(2.4.3)$$

Thus if $\mathrm{d}x/\mathrm{d}t = v + \lambda H$, the first two terms on the right of (2.4.3) give

$$\frac{\mathrm{d}v}{\mathrm{d}t} = \frac{\partial v}{\partial x} \frac{\mathrm{d}x}{\mathrm{d}t} + \frac{\partial v}{\partial t} \qquad (2.4.4)$$

and if $\mathrm{d}x/\mathrm{d}t = v + (g/\lambda)$, the third term on the right of (2.4.3) has the main curly bracket equal to

$$\frac{\mathrm{d}H}{\mathrm{d}t} = \frac{\partial H}{\partial x} \frac{\mathrm{d}x}{\mathrm{d}t} + \frac{\partial H}{\partial t}. \qquad (2.4.5)$$

Then eqn (2.4.3) gives

$$\lambda H_1 + H_2 = \frac{\mathrm{d}v}{\mathrm{d}t} + \lambda \frac{\mathrm{d}H}{\mathrm{d}t} + g(s_f - s) = 0. \qquad (2.4.6)$$

The two expressions for $\mathrm{d}x/\mathrm{d}t$ are the same if $v + \lambda H = v + (g/\lambda)$, i.e. $\lambda = \pm \sqrt{(g/H)}$. These two values of λ give the characteristics which are similar to those shown in Fig. 2.3.2. $\mathrm{d}x/\mathrm{d}t = v + \sqrt{(gH)}$ is the characteristic PR with positive slope. Since the flow is taken to be subcritical so that $v < \sqrt{(gH)}$, the characteristic QR given by $\mathrm{d}x/\mathrm{d}t = v - \sqrt{(gH)}$ has negative slope, otherwise this method would not work.

The St Venant equations can now be represented in increment form as

$$\Delta v + \Delta H \sqrt{(g/H)} + g(s_f - s)\Delta t = 0 \qquad (2.4.7)$$

on

$$\Delta x = \{v + \sqrt{(gH)}\} \Delta t \qquad (2.4.8)$$

and

$$\Delta v - \Delta H \sqrt{(g/H)} + g(s_f - s) = 0 \qquad (2.4.9)$$

on

$$\Delta x = \{v - \sqrt{(gH)}\} \Delta t. \qquad (2.4.10)$$

It is supposed that the initial values of v and H at $t = 0$ are given. Then with the increments approximated by differences the solution can be stepped out along the characteristic curves as follows.

Put $f = v + \sqrt{(gH)}$ and $g = v - \sqrt{(gH)}$. Suppose that the values of H and v at points P, Q (not on the same characteristic) are known. Then eqns (2.4.7) to (2.4.10) are approximated by

$$v_R - v_P + (H_R - H_P)\sqrt{(g/H_P)} + g(s_{fP} - s)(t_R - t_P) = 0 \qquad (2.4.11)$$

$$x_R - x_P = f_P(t_R - t_P) \qquad (2.4.12)$$

$$v_R - v_Q - (H_R - H_Q)\sqrt{(g/H_Q)} + g(s_{fQ} - s)(t_R - t_Q) = 0 \qquad (2.4.13)$$

$$x_R - x_Q = g_Q(t_R - t_Q) \qquad (2.4.14)$$

where s_{fP}, s_{fQ} are the friction terms evaluated with the known values at P, Q respectively. This is a set of four simultaneous equations for the four unknowns x_R, t_R, H_R, v_R. The values of x_R and t_R can be obtained from eqns (2.4.12) and (2.4.14) as

$$t_R = \frac{x_Q - x_P + f_P t_P - g_Q t_Q}{f_P - g_Q} \qquad (2.4.15)$$

and

$$x_R = x_P + f_P(t_R - t_P). \qquad (2.4.16)$$

Subtract (2.4.13) from (2.4.11) to eliminate v_R and this gives H_R from

$$H_R\{\sqrt{(g/H_P)} + \sqrt{(g/H_Q)}\} = v_P - v_Q + \sqrt{(gH_P)}$$
$$+ \sqrt{(gH_Q)} - g(s_{fP} - s)(t_R - t_P)$$
$$+ g(s_{fQ} - s)(t_R - t_Q). \qquad (2.4.17)$$

This value of H_R can then be substituted into (2.4.11) to give

$$v_R = v_P - (H_R - H_P)\sqrt{(g/H_P)} - g(s_{fP} - s)(t_R - t_P). \qquad (2.4.18)$$

This first approximation for H_R and v_R amounts to assuming that the arcs PR and QR are straight lines. The approximation can be improved by repeating the calculation using averaged values for the non-linear terms, i.e. $\sqrt{(g/H_P)}$ replaced by $\frac{1}{2}[\sqrt{(g/H_P)} + \sqrt{(g/H_R)}]$ and s_{fP} replaced by $\frac{1}{2}(s_{fP} + s_{fR})$ in (2.4.11), f_P replaced by $\frac{1}{2}(f_P + f_R)$ in (2.4.12) and $\sqrt{(g/H_Q)}$ replaced by $\frac{1}{2}[\sqrt{(g/H_Q)} + \sqrt{(g/H_R)}]$, and s_{fQ} replaced by $\frac{1}{2}(s_{fQ} + s_{fR})$ in (2.4.13) and g_Q replaced by $\frac{1}{2}(g_Q + g_R)$ in (2.4.14).

The process is repeated until the numerical values repeat to a required number of figures. If convergence is not obtained in a small number of iterations it is a sign that the points P and Q have been taken too far apart.

The example in section 1.5 of the effect of rainfall on a slope illustrates the way the characteristics 'carry' information about the solution. In Figure 1.5.1 the characteristic $x = c(t - t_s)$ carries the information that at time $t = t_s$ it had stopped raining. Note that it is not possible to calculate the depth of water on the slope at points in the x, t space above this characteristic without information as to what happens to the rainfall when $t > t_s$. This is equivalent to saying that the numerical solution must not attempt to step out beyond the characteristic through the furthest upstream point from which information is being used to obtain this solution. This is the essence of the CFL (Courant-Friedrichs-Lewy, 1928, 1967) condition which is closely associated with the conditions for stability of finite difference schemes and is discussed further in section 3.10.

Further points to consider are:

1. The surface flow equations for which the method of characteristics is used in this chapter are obtained from the basic equations of continuity and momentum as shown in section 1.4. The satisfaction of the equation of continuity guarantees conservation of the mass of water. The satisfaction of the equation of momentum guarantees the balance:

loss of (kinetic plus potential) energy = work done against friction

$$(2.4.19)$$

The momentum equation is obtained by differentiating this. When a numerical scheme is used so that these equations are only satisfied approximately there is likely to be an imbalance. An unacceptable discrepancy in the conservation of mass is another useful indicator that more refinement is needed, e.g. the points P, Q in Figure 2.3.2 should be closer.

2. Another consideration is whether the one-dimensional channel equations in section 1.4 are an adequate representation of the problem. If the channel is actually of an irregular shape and alignment then oblique waves can be generated and persist downstream. Ellis and Pender (1982) describe a numerical model for the analysis of flow in a chute spillway using the numerical method of characteristics in two-dimensional flow (Abbott, 1979) plus interpolation onto a fixed grid. This avoids a major disadvantage of the numerical method of characteristics that it gives the solution at places and time which cannot be fixed beforehand. Other examples of this approach are given in Chapter 9. The method of finite differences described in Chapter 3 obtains the solution directly at points on a fixed grid.

3 Finite difference methods

3.1 Introduction

In this chapter the finite difference method is presented in a general way suitable for the applications in this book. The method is mainly illustrated on dimensionless equations which have the same form as those in the surface and subsurface flow problems which are the subject of this book. The basic idea is that the derivatives in the differential equations which represent the water resources problems are approximated by expressions using differences between the values of the dependent variable at selected points of a grid. The differential equation is thus replaced by a number of algebraic equations. The finite difference method is first introduced with the simplest replacements for the derivatives. The more sophisticated methods which are also included in this chapter are all in current use in the solution of water resource problems. The references for these are given in later chapters. Readers who are not interested in convergence proofs can skip them but it is essential to know about stability (sections 3.3, 3.7, 3.10).

In Chapter 1 it is shown that the differential equations can be separated into categories: ordinary and partial differential equations and each of these into initial and boundary value problems. The following sections discuss these separately.

3.2 Ordinary differential equations—initial value problems

The finite difference methods here are illustrated as applied to the general equation

$$\frac{dy}{dt} = f(y, t), \quad t > 0 \tag{3.2.1}$$

with given initial value $y(0)$. Here $y(t)$ could be the concentration of a pollutant which grows according to the law given by (3.2.1). The simplest finite difference method for this equation is Euler's method which is equivalent to stepping out along the tangent as shown in Fig. 3.2.1. The slope of the tangent at $t = 0$ is

$$\tan \psi_0 = f_0 \tag{3.2.2}$$

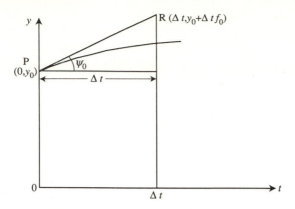

Figure 3.2.1: Euler approximation.

where $f_0 = f(y_0, 0) = (dy/dt)_0$, from (3.2.1). Thus Euler's method approximates the differential equation with

$$\frac{y_{n+1} - y_n}{\Delta t} - f_n = 0, \quad n = 0, 1, 2, \ldots \tag{3.2.3}$$

which is equivalent to

$$y_{n+1} = y_n + \Delta t f_n \tag{3.2.4}$$

with $y_0 = y(0)$. This is exact for a straight line but not for a curve. The error can be obtained by using a Taylor series expansion (Conte and de Boor, 1980). First the difference scheme is written in the form (3.2.3) in which it is a direct replacement of the differential equation written with zero on the right-hand side so that it represents the result of a difference operator acting on y_n:

$$L_{\Delta t}(y_n) = \frac{1}{\Delta t}(y_{n+1} - y_n) - f_n = 0. \tag{3.2.5}$$

Then the local truncation error τ_n is obtained by looking at the effect of the difference operator in (3.2.5) when it acts on the exact solution of the differential equation:

$$\tau_n = L_{\Delta t}(y(n\Delta t)) = \frac{1}{\Delta t}[y((n+1)\Delta t) - y(n\Delta t)] - f(y(n\Delta t), n\Delta t). \tag{3.2.6}$$

The right-hand side of (3.2.6) is expanded by Taylor series and the error due to truncating this gives the local truncation error. Using y and its derivatives as at $t = n\Delta t$ and using the differential equation (3.2.1) then gives

$$\tau_n = \frac{1}{\Delta t}\left[\left(y + \Delta t\,\frac{dy}{dt} + \frac{\Delta t^2}{2}\frac{d^2y}{dt^2} + \cdots\right) - y\right] - \frac{dy}{dt}$$

$$= \frac{\Delta t}{2}\frac{d^2y}{dt^2} + \text{higher powers of } \Delta t. \tag{3.2.7}$$

The truncation error with the Euler scheme thus involves the second derivative of y as expected. In the application of the numerical method the equation should be non-dimensionalized and the time increment expressed as a fraction of the total time of the event being studied; thus Δt must certainly be less than unity. The general process of non-dimensionalizing is discussed in Chapter 10. The smaller Δt is the better so that the truncation error remains small. The truncation error in (3.2.7) is said to be of first order in Δt and this is written $O(\Delta t)$. The statement 'error is $O(\Delta t)$' means that 'the ratio error/Δt tends to a constant value as Δt tends to zero'. This will not necessarily be apparent for larger values of Δt but only as Δt becomes small enough. When the local truncation error tends to zero with Δt the scheme is said to be consistent.

A more accurate method is obtained by integrating eqn (3.2.1) over the time interval from $n\Delta t$ to $(n + 1)\Delta t$:

$$y((n + 1)\Delta t) - y(n\Delta t) = \int_{n\Delta t}^{(n + 1)\Delta t} f(y, t)\,dt$$

and approximating this using the trapezium rule (Crank and Nicolson, 1949):

$$y_{n+1} - y_n = \frac{\Delta t}{2}(f_{n+1} + f_n). \tag{3.2.8}$$

This effectively uses the average of the slopes of the tangents to the curve at $n\Delta t$ and $(n + 1)\Delta t$. The truncation error found in exactly the same way as described above for the Euler method is now given by

$$\tau_n = \frac{1}{\Delta t}\left[y + \Delta t\,\frac{dy}{dt} + \frac{\Delta t^2}{2}\frac{d^2y}{dt^2} + \frac{\Delta t^3}{6}\frac{d^3y}{dt^3} + \cdots - y\right]$$

$$- \frac{1}{2}\left[\frac{dy}{dt} + \Delta t\,\frac{d^2y}{dt^2} + \frac{\Delta t^2}{2}\frac{d^3y}{dt^3} + \cdots + \frac{dy}{dt}\right]$$

$$= -\frac{\Delta t^2}{12}\frac{d^3y}{dy^3} + \text{higher powers of } \Delta t. \tag{3.2.9}$$

Thus the trapezium rule scheme is consistent and has error $O(\Delta t^2)$;

hence it is more accurate (since Δt is essentially less than unity, the higher the power of Δt the smaller the result).

The trapezium rule is more accurate than Euler but in some kinds of problem it can have disadvantages as explained later. It is frequently considered better to use the backward difference formula

$$\frac{y_{n+1} - y_n}{\Delta t} = f_{n+1} \qquad (3.2.10)$$

even though it has only a first-order error (the result is the same as (3.2.7)) because it produces a smoother result.

The Euler method is called explicit because eqn (3.2.4) gives a rule for obtaining y_{n+1} explicitly. The trapezium rule (3.2.8) and the backward difference scheme (3.2.10) are implicit because the unknown y_{n+1} is in f_{n+1} also. This is simple if $f(y, t)$ is linear in y but if not a predictor–corrector method can be used. With this the Euler scheme is used for a first approximation $y_{n+1}^{(1)}$, the 'predictor', and then the trapezium rule can be used as a 'corrector' to improve the approximation successively:

$$y_{n+1}^{(k+1)} = y_n + \frac{\Delta t}{2} (f_n + f_{n+1}^{(k)})$$

where $f_{n+1}^{(k)} = f(y_{n+1}^{(k)}, (n+1)\Delta t)$. The method can be referred to as PC^m if the corrector is applied m times. The method PC with the corrector applied once can be illustrated by the equation

$$\frac{dy}{dt} = -\lambda y, \qquad \lambda \text{ constant.} \qquad (3.2.11)$$

The predictor gives

$$y_{n+1}^{(1)} = (1 - \lambda \Delta t) y_n$$

and the corrector gives

$$y_{n+1}^{(2)} = y_n - \lambda \frac{\Delta t}{2} (y_n + y_{n+1}^{(1)}).$$

Thus the net effect is that the differential equation is approximated by

$$L_{\Delta t}(y_n) = \frac{y_{n+1}^{(2)} - y_n}{\Delta t} + \lambda y_n - \frac{\lambda^2 \Delta t}{2} y_n = 0. \qquad (3.2.12)$$

The local truncation error obtained by the same method as in (3.2.6) is given by

$$\tau = L_{\Delta t}(y(n\Delta t)) = \frac{dy}{dt} + \lambda y + \frac{\Delta t}{2} \left(\frac{d^2 y}{dt^2} - \lambda^2 y \right) + O(\Delta t^2).$$

Using the differential equation (3.2.11) and the result of differentiating
it shows that the local truncation error is $O(\Delta t^2)$. Thus the truncation
error of the result is the same as that of the corrector, i.e. the corrector
dominates the result. When the function $f(y, t)$ is non-linear the
corrector step can be repeated until two successive iterates differ by
less than some required tolerance. This is called iterating to convergence. When the function $f(y, t)$ can be differentiated easily with
respect to y a much faster iteration method is Newton's method
described in Appendix 2.

   The size of the time step Δt has to be controlled in order to keep
down the size of the truncation error. Another property of time-
stepping schemes which controls the time step size is that of stability
and this is discussed in the next section.

3.3 Stability

Suppose the problem to be solved is represented by the equation

$$\frac{dy}{dt} = -\lambda y \qquad (3.3.1)$$

where λ is a positive constant. The solution of this equation is
known to be $y = y(0) \exp(-\lambda t)$ where $y(0)$ is the initial value at $t = 0$.
It is used here as an example of an equation representing a problem
where it is known from the physics that the quantity $y(t)$ must decay
(e.g. an effluent which is biodegradable). The Euler scheme (3.2.4) for
solving (3.3.1) gives

$$y_{n+1} = (1 - \lambda \Delta t)y_n. \qquad (3.3.2)$$

The factor $\xi = y_{n+1}/y_n$ which muliplies the numerical solution at each
time step is called the amplification factor. Since the exact solution
decays it is essential that the numerical solution decays also. The
numerical scheme is then said to be stable. For this it is necessary
that the amplification factor should have modulus (numerical value)
less than unity, i.e.

$$|\xi| = |1 - \lambda \Delta t| \le 1. \qquad (3.3.3)$$

Since $1 - \lambda \Delta t < 1$ the condition is required for $1 - \lambda \Delta t > -1$. Hence
for stability of the Euler scheme it is required that $\lambda \Delta t < 2$, i.e.

$$\Delta t < 2/\lambda. \qquad (3.3.4)$$

In problems where the value of λ can be very large this restriction
makes the Euler scheme uneconomic.

The trapezium rule (3.2.8) applied to the solution of eqn (3.3.1) gives

$$\frac{y_{n+1} - y_n}{\Delta t} = -\frac{\lambda}{2}(y_{n+1} + y_n)$$

i.e.

$$y_{n+1} = \frac{(1 - \frac{1}{2}\lambda\Delta t)}{(1 + \frac{1}{2}\lambda\Delta t)} y_n. \tag{3.3.5}$$

Recall that the parameter λ is positive and hence the amplification factor here has modulus less than unity if

$$|1 - \tfrac{1}{2}\lambda\Delta t| < 1 + \tfrac{1}{2}\lambda\Delta t$$

i.e.

$$\tfrac{1}{2}\lambda\Delta t - 1 < 1 + \tfrac{1}{2}\lambda\Delta t$$

which is true for any value of Δt. Hence the trapezium rule scheme is said to be unconditionally stable, i.e. there is no condition on the value of the time step Δt. Similarly the backward difference method (3.2.10) used on the equation (3.3.1) has amplification factor

$$\xi = 1/(1 + \lambda\Delta t) \tag{3.3.6}$$

which is again less than unity for any value of Δt and the backward difference scheme is unconditionally stable. In general the more implicit methods are more stable.

The trapezium rule scheme (3.2.8) has a higher-order truncation error because it is effectively centralized about the $(n + \frac{1}{2}\Delta t)$ time level. It is important to realize that centralizing a difference scheme to get this effect is not always a good idea. The notorious example of this is the midpoint rule which replaces eqn (3.2.1) with the approximation

$$\frac{y_{n+2} - y_n}{2\Delta t} = f_{n+1}. \tag{3.3.7}$$

Applying this to (3.3.1) gives

$$y_{n+2} + 2\lambda\Delta t y_{n+1} - y_n = 0. \tag{3.3.8}$$

This is a difference equation (or recurrence relation) in y_n which can be solved by putting $y_n = r^n$ (Conte and de Boor, 1980); if r_1 and r_2 are the roots of the quadratic characteristic polynomial

$$r^2 + 2\lambda\Delta t r - 1 = 0 \tag{3.3.9}$$

the general solution is $y_n = Ar_1^n + Br_2^n$ where A and B depend on the initial conditions. But eqn (3.3.9) has real roots and their product is

-1. Hence one root is numerically greater than unity and the scheme is unconditionally unstable.

The scheme (3.3.8) needs two starting values: $y_0 = y(0)$ but y_1 must be separately generated. Even if y_1 is chosen to be $r_1 y(0)$ where r_1 is the numerically smaller root of (3.3.9), rounding errors will bring in the other root which will dominate the result. The inexperienced frequently attempt to use this method with parabolic problems (section 3.7). It was originally proposed by Richardson (1910). Richardson thought that all difference expressions should be centralized but his calculations did not get far enough to reveal what was wrong with the scheme. However, the midpoint rule, (3.3.7), is useful in connection with methods which use Richardson extrapolation (Richardson, 1927).

Extrapolation methods

These lead to an algorithm for the solution of a system of equations representing the interaction of various solutes in water-quality modelling (Chapter 9). If an approximate solution of (3.2.1) using a time step h is known to give a value

$$A(h) = y(h) + A_1 h + A_2 h^2 + \cdots \text{ higher powers of } h \quad (3.3.10)$$

where $A_1, A_2, \ldots$ are independent of h, then using a time step $h/2$ gives the approximate solution

$$A\left(\frac{h}{2}\right) = y(h) + A_1 \frac{h}{2} + A_2 \frac{h^2}{4} + \cdots. \quad (3.3.11)$$

Combining these as

$$2A\left(\frac{h}{2}\right) - A(h) = y(h) - \tfrac{1}{2}A_2 h^2 + \cdots \quad (3.3.12)$$

gives an improved result with error now $O(h^2)$. This is the basis of Richardson extrapolation. Further, if $A(h/4)$ is also computed a linear combination of the three values to eliminate A_2 also will give a result with error $O(h^3)$.

The method is more effective if the algorithm is such that (3.3.10) is replaced by

$$A(h) = y(h) + A_2 h^2 + A_4 h^4 + \cdots \quad (3.3.13)$$

i.e. only even powers. Then the extrapolation sends the error from $O(h^2)$ to $O(h^4)$ and to $O(h^6)$ with more values combined. Equation (3.3.12) becomes

$$\tfrac{4}{3}A\left(\frac{h}{2}\right) - \tfrac{1}{3}A(h) = y(h) + O(h^4). \quad (3.3.14)$$

The midpoint rule (3.3.7), with a starting value y_1 given by the Euler method (3.2.3), gives an error of the form (3.3.13) provided that the number of times the method is applied is always even or always odd (Gragg, 1965). In Gragg's method (also called the modified midpoint method), the possible instability of the midpoint rule is controlled by a smoothing step and is defined as follows.

To calculate $y(\Delta t)$, the solution of (3.2.1), given $y(0)$: $h_s = \Delta t/N_s$, N_s is an even integer (the time step Δt is divided into N_s sub-time steps)

$$y_0 = y(0)$$

$$y_1 = y_0 + h_s f(0, y_0) \quad \text{(Euler start)}$$

$$y_{m+2} - y_m = 2h_s f((m+1)h_s, y_{m+1}), \quad m = 0, 1, 2, \ldots, N_s - 1$$

(midpoint rule applied N_s times)

$$y(\Delta t; h_s) = \tfrac{1}{4}y_{N_s+1} + \tfrac{1}{2}y_{N_s} + \tfrac{1}{4}y_{N_s-1} \quad \text{(smoothing)} \qquad (3.3.15)$$

where the left-hand side of (3.3.15) means the value of $y(\Delta t)$ calculated with sub-time steps h_s. The sequence $\{N_s\}$ is frequently chosen as $\{2, 4, 6, 8, 12, 16, 24, \ldots\}$ or $\{2, 4, 8, 16, 32, 64, \ldots\}$.

The Richardson extrapolation applied to Gragg's method gives a series of terms $a_s^{(m)}$ defined as follows: for each value of h_s compute $A(h_s) = a_s^{(0)}$; combining $A(h_0)$ and $A(h_1)$ gives $a_0^{(1)}$, error $O(h_0^4)$; combining $A(h_1)$ and $A(h_2)$ gives $a_1^{(1)}$, error $O(h_1^4)$; combining $A(h_0)$, $A(h_1)$, and $A(h_2)$ gives $a_0^{(2)}$, error $O(h_0^6)$, etc.; and the polynomial extrapolation is given by the recurrence

$$a_s^{(0)} = A(h_s), \quad a_s^{(m)} = a_{s+1}^{(m-1)} + \frac{a_{s+1}^{(m-1)} - a_s^{(m-1)}}{(h_s/h_{m+s})^2 - 1},$$

$$m = 1, 2, \ldots, s = 0, 1, 2, \ldots.$$

The result is

$$a_s^{(m)} = y(\Delta t) + O(h_s^{2m+2}). \qquad (3.3.16)$$

Details are given in Lambert (1973). Burden *et al.* (1981) give the algorithm with step size control.

Instead of fitting a polynomial for the extrapolation the method of Bulirsch and Stoer (1966) uses rational functions of the form $P(h)/Q(h)$ where $P(h)$ and $Q(h)$ are polynomials. Applied to Gragg's modified midpoint method this is known as the GBS scheme. It is declared by Hull *et al.* (1971) to be the best method to use when the function evaluations are relatively inexpensive. It can also be used with a step control procedure.

Runge–Kutta methods

The extrapolation method is a kind of one-step method with division into substeps. Runge–Kutta methods (Wood, 1990) are also one-step methods which are useful for the solution of a system of ordinary differential equations of the form

$$\frac{d\mathbf{C}}{dt} = \mathbf{g}(t, \mathbf{C}) \qquad (3.3.17)$$

where the vector $\mathbf{C} = (C_1, C_2, \ldots)^{\mathrm{T}}$ can represent the concentrations of various dissolved pollutants in surface or groundwater flow and eqn (3.3.17) expresses their chemical reactions with each other. Since they are one-step methods it is easy to change the time step size and there are no spurious solutions. The most popular Runge–Kutta method is a fourth-order method, called this because the cumulative error is $O(\Delta t^4)$. It is sometimes called the fourth-order Runge–Kutta method but there are others. This method is given by

$$\mathbf{C}_{n+1} = \mathbf{C}_n + \Delta t(\mathbf{k}_1 + 2\mathbf{k}_2 + 2\mathbf{k}_3 + \mathbf{k}_4)/6 \qquad (3.3.18)$$

where

$$\mathbf{k}_1 = \mathbf{g}(t_n, \mathbf{C}_n)$$

$$\mathbf{k}_2 = \mathbf{g}(t_n + 0.5\Delta t, \mathbf{C}_n + 0.5\Delta t\mathbf{k}_1)$$

$$\mathbf{k}_3 = \mathbf{g}(t_n + 0.5\Delta t, \mathbf{C}_n + 0.5\Delta t\mathbf{k}_2)$$

$$\mathbf{k}_4 = \mathbf{g}(t_n + \Delta t, \mathbf{C}_n + \Delta t\mathbf{k}_3).$$

The $\mathbf{k}_1, \mathbf{k}_2, \ldots$ are formed recursively and the method is very simple to implement.

Note that the Euler, trapezium rule, and backward difference schemes are all examples of the θ method:

$$\frac{y_{n+1} - y_n}{\Delta t} = \theta f_{n+1} + (1 - \theta)f_n, \quad 0 \le \theta \le 1 \qquad (3.3.19)$$

with $\theta = 0, 0.5$, and 1 respectively. Analysing the stability of the θ method as applied to (3.3.1) in the same way as above, the amplification factor is

$$\xi = \frac{1 - (1 - \theta)\lambda\Delta t}{1 + \theta\lambda\Delta t}. \qquad (3.3.20)$$

The stability condition $|\xi| < 1$ gives

$$(1 - \theta)\lambda\Delta t - 1 < 1 + \theta\lambda\Delta t$$

i.e.

$$(1 - 2\theta)\lambda\Delta t < 2. \tag{3.3.21}$$

For $\theta \geq 0.5$ condition (3.3.21) is satisfied for all Δt and hence there is unconditional stability. For $\theta < 0.5$ the stability condition is

$$\lambda\Delta t < \frac{2}{1 - 2\theta}. \tag{3.3.22}$$

These last two results include the special cases $\theta = 0, 0.5, 1$ discussed above.

The exact solution of (3.3.1) is multiplied by $\exp(-\lambda\Delta t)$ at each time step. The amplification factor ξ is approximating this. If $\lambda\Delta t$ is small, expanding the right-hand side of (3.3.20) in powers of $\lambda\Delta t$ gives

$$\xi = (1 + \theta\lambda\Delta t)^{-1}[1 - (1 - \theta)\lambda\Delta t]$$

i.e.

$$\xi = 1 - \lambda\Delta t + \theta\lambda^2\Delta t^2 + \cdots. \tag{3.3.23}$$

Because ξ here is approximating $\exp(-\lambda\Delta t)$ in the exact solution of (3.3.1) this shows the extra accuracy given by $\theta = 0.5$.

The scheme (3.3.7) or (3.3.8) is a two-step scheme (because it involves y_n, y_{n+1}, y_{n+2}) for the solution of a first-order equation. This means that the quadratic (3.3.9) has one root which is approximating $\exp(-\lambda\Delta t)$:

$$r_1 = -\lambda\Delta t + \sqrt{(1 + \lambda^2\Delta t^2)}$$
$$= 1 - \lambda\Delta t + \tfrac{1}{2}\lambda^2\Delta t^2 + \cdots$$

and the other root, r_2, which is spurious. This spurious root is introduced by the numerical scheme and in this case it is parasitic, i.e. it grows and swamps the root r_1.

Another scheme which is best avoided because of trouble with a spurious root is Lees' algorithm, (Lees, 1966). This seems an attractive proposition for dealing with a non-linear form:

$$\frac{dy}{dt} = -\lambda(y)y. \tag{3.3.24}$$

The algorithm represents this equation as

$$\frac{y_{n+1} - y_{n-1}}{2\Delta t} + \tfrac{1}{3}\lambda(y_n)(y_{n+1} + y_n + y_{n-1}) = 0. \tag{3.3.25}$$

The stability can be analysed by taking the linearized form and putting $y_n = r^n$ as in (3.3.8) to give

$$(1 + \tfrac{2}{3}\lambda\Delta t)r^2 + \tfrac{2}{3}\lambda\Delta t r + (-1 + \tfrac{2}{3}\lambda\Delta t) = 0. \tag{3.3.26}$$

The principal root is

$$r_1 = [-\tfrac{1}{3}\lambda\Delta t + \sqrt{(1 - \tfrac{1}{3}\Delta t^2 \lambda^2)}]/(1 + \tfrac{2}{3}\lambda\Delta t) \qquad (3.3.27)$$

which expands into

$$r_1 = 1 - \lambda\Delta t + \tfrac{1}{2}\lambda^2\Delta t^2 + O(\Delta t^3). \qquad (3.3.28)$$

This is a good approximation to $\exp(-\lambda\Delta t)$ but the spurious root, r_2, is always numerically larger and negative and there will always be trouble with oscillations (Wood, 1990). Bettencourt *et al.* (1981) call Lees' algorithm 'notoriously oscillatory' but it is still being considered as an option in 1990 (Paniconi *et al.*). This book only advocates schemes which avoid any difficulties with spurious roots.

In the next section the general θ method form is used in the discussion of the important concept of convergence.

3.4 Convergence

This section includes a simplified version of the Lax equivalence theorem which says that if a difference scheme for the solution of a (well-posed) initial value problem is consistent and stable then as the time step Δt is made smaller the solution of the difference scheme will converge to the solution of the differential equation. (The convergence discussed here is not to be confused with the convergence of the iteration necessary in the method of characteristics (Chapter 2) or in solving a non-linear equation as discussed in Appendix 2.) This theorem is demonstrated here on the simple differential equation,

$$\frac{dy}{dt} = -\lambda y, \qquad \lambda \text{ constant} \qquad (3.4.1)$$

corresponding to a physical situation where the quantity *y(t)* decays.

Suppose the θ method (3.3.19) applied to the differential equation (3.4.1) is written as

$$L_{\Delta t}(y_n) = 0. \qquad (3.4.2)$$

Then as before the local truncation error is given by

$$L_{\Delta t}(y(n\Delta t)) = \tau_n. \qquad (3.4.3)$$

Subtract (3.4.2) from (3.4.3) and put $e_n = y(n\Delta t) - y_n$. The result is

$$\frac{1}{\Delta t}(e_{n+1} - e_n) + \lambda[\theta e_{n+1} + (1 - \theta)e_n] = \tau_n. \qquad (3.4.4)$$

Write this as

$$e_{n+1} = \xi e_n + \beta_n. \qquad (3.4.5)$$

Then

$$\xi = [1 - \lambda(1 - \theta)\Delta t]/(1 + \lambda\theta\Delta t) \qquad (3.4.6)$$

is the amplification factor and $\beta_n = \Delta t\tau_n/(1 + \lambda\theta\Delta t)$. Equation (3.4.5) is a difference equation in e_n which can be solved by writing

$$e_n = \xi e_{n-1} + \beta_{n-1}$$
$$= \xi(\xi e_{n-2} + \beta_{n-2}) + \beta_{n-1}$$
$$= \cdots$$

i.e.

$$e_n = \xi^n e_0 + \xi^{n-1}\beta_0 + \cdots + \xi\beta_{n-2} + \beta_{n-1}.$$

Hence it follows that

$$|e_n| \le |\xi|^n|e_0| + |\xi|^{n-1}|\beta_0| + \cdots + |\xi||\beta_{n-2}| + |\beta_{n-1}| \qquad (3.4.7)$$

If the scheme is stable the modulus of the amplification factor $|\xi|$ is less than or equal to unity and if $\tau = \max_n|\tau_n|$, (3.4.7) leads to

$$|e_n| \le |e_0| + \frac{n\Delta t}{1 + \lambda\theta\Delta t}\tau. \qquad (3.4.8)$$

Thus for a fixed time $T = n\Delta t$ and supposing that the starting error is zero, the modulus of the error in the solution is

$$|e_n| \le T\tau$$

and if the scheme is consistent τ certainly tends to zero as the time step size decreases; hence the numerical solution converges to the solution of the differential equation.

The disadvantage of the trapezium rule (Crank–Nicolson) scheme mentioned earlier is that because it is unconditionally stable large values of Δt may be used thus giving an inaccurate solution. If also the problem being solved involves large values of the parameter λ, the amplification factor seen in (3.3.5) is approximately -1. As seen in (3.4.5), the error is multiplied by this amplification factor and this can produce the sawtooth effect shown in Fig. 3.4.1 when the solution should be smooth (Wood and Lewis, 1975). Because this effect is associated with the method proposed by Crank and Nicolson (1949) it is referred to as the Crank–Nicolson noise effect (Wood, 1990).

In section 3.3 it is shown that the amplification factor is supposed to be approximating $\exp(-\lambda\Delta t)$ which is positive. From (3.4.6) it can be seen that the amplification factor ξ is negative when

$$(1 - \theta)\lambda\Delta t > 1$$

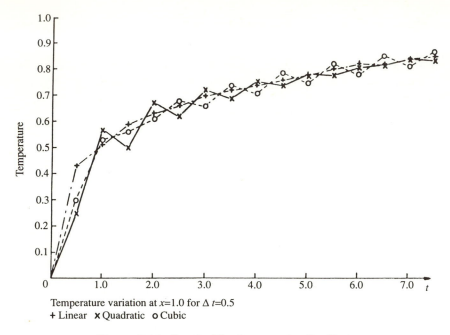

Temperature variation at x=1.0 for $\Delta\,t$=0.5

$+$ Linear $\times$ Quadratic   o Cubic

Figure 3.4.1: Crank–Nicolson sawtooth effect.

so that any value of θ except $\theta = 1$ is liable to give a ripple effect on what should be a smooth solution if large values of $\lambda \Delta t$ occur.

3.5 Ordinary differential equations—boundary value problems

An example of an ordinary differential equation representing a boundary value problem is

$$-\frac{d^2 u}{dx^2} + qu = f(x), \quad q \geq 0 \tag{3.5.1}$$

in a region $a \leq x \leq b$, with information given at the boundary points such as

$$u(a) = \alpha \tag{3.5.2}$$

and

$$\frac{du}{dx} + \beta u = \gamma, \quad \text{at } x = b. \tag{3.5.3}$$

Equation (3.5.2) is called a Dirichlet-type boundary condition. If $\beta = 0$, then (3.5.3) is the Neumann boundary condition (usually representing a known flow)

$$\frac{du}{dx} = \gamma. \tag{3.5.4}$$

Otherwise (3.5.3) is a mixed-type boundary condition.

Equation (3.5.1) is a simplified form of the equation for the lateral distribution of velocity across a channel discussed in section 4.1.

The region $[a, b]$ is divided into n equal subintervals of length Δx, i.e. $b - a = n\Delta x$. The second derivative is always approximated at a general grid point $x = j\Delta x$ by the centralized finite difference expression

$$\frac{d^2 u}{dx^2} \approx \frac{u_{j-1} - 2u_j + u_{j+1}}{\Delta x^2}. \tag{3.5.5}$$

Thus the differential equation (3.5.1) is approximated by the set of finite difference equations

$$L_{\Delta t}(u_j) = \frac{-u_{j-1} + 2u_j - u_{j+1}}{\Delta x^2} + qu_j - f_j = 0, \quad j = 1, 2, \ldots, n \tag{3.5.6}$$

supposing q is a constant. The typical equation of the set (3.5.6) is written here as the result of a difference operator $L_{\Delta x}$ acting on u_j and with zero on the right-hand side so that the local truncation error of the scheme, τ_j, can be found in the same way as in section 3.2 by applying the same difference operator to the exact solution $u(j\Delta x)$:

$$\tau_j = L_{\Delta x}(u(j\Delta x)). \tag{3.5.7}$$

Expanding by Taylor series (Conte and de Boor, 1980) gives

$$\tau_j = -\frac{1}{\Delta x^2}\left[u - \Delta x \frac{du}{dx} + \frac{\Delta x^2}{2}\frac{d^2 u}{dx^2} - \frac{\Delta x^3}{6}\frac{d^3 u}{dx^3} + \frac{\Delta x^4}{24}\frac{d^4 u}{dx^4} + \cdots - 2u + u \right.$$

$$\left. + \Delta x \frac{du}{dx} + \frac{\Delta x^2}{2}\frac{d^2 u}{dx^2} + \frac{\Delta x^3}{6}\frac{d^3 u}{dx^3} + \frac{\Delta x^4}{24}\frac{d^4 u}{dx^4} + \cdots \right] + qu - f$$

$$= -\frac{d^2 u}{dx^2} + qu - f - \frac{\Delta x^2}{12}\frac{d^4 u}{dx^4}$$

$$= -\frac{\Delta x^2}{12}\frac{d^4 u}{dx^4}$$

using (3.5.1). The local truncation error τ is $O(\Delta x^2)$ as expected with a centralized scheme and the scheme is consistent.

The differential equation (3.5.1) is thus represented by the set of finite difference equations (3.5.6) for the $(n+1)$ unknowns $u_1, u_2, \ldots, u_n, u_{n+1}$ where u_{n+1} is a fictitious value at $x = b + \Delta x$ which is to be included in the difference representation of the boundary condition (3.5.3), to complete the set of $(n+1)$ equations. There are various ways of doing this. A centralized form is

$$\frac{u_{n+1} - u_{n-1}}{2\Delta x} + \beta u_n = \gamma. \tag{3.5.8}$$

This will have the same order, i.e. $O(\Delta x^2)$, as the other equations of the set (as is seen in section 3.2) because it is centralized. The equation (3.5.8) can be either used to eliminate the unwanted u_{n+1} from the last equation of the set (3.5.6) or tacked on to complete the set. The first equation of the set (3.5.6) includes the Dirichlet boundary condition at $x = a$ by writing it as

$$\left(\frac{2}{\Delta x^2} + q\right)u_1 - \frac{u_2}{\Delta x^2} = f_1 + \frac{\alpha}{\Delta x^2}. \tag{3.5.9}$$

The result is a set of simultaneous linear equations

$$\mathbf{A}\mathbf{u} = \mathbf{f} \tag{3.5.10}$$

for the values of u at the grid points. The matrix $\mathbf{A}$ is tridiagonal because each equation has no more than three unknowns. Hence the system can be solved by the standard tridiagonal matrix solver (see Conte and de Boor, 1980), (this is sometimes called the Thomas algorithm).

If the ordinary differential equation boundary value problem also includes a first-derivative term as in

$$-\frac{d^2u}{dx^2} + p\frac{du}{dx} + qu = f \tag{3.5.11}$$

this first derivative can be approximated by the central difference formula

$$\frac{du}{dx} \approx \frac{u_{j+1} - u_{j-1}}{2\Delta x}. \tag{3.5.12}$$

This will preserve the truncation error at $O(\Delta x^2)$. The matrix $\mathbf{A}$ is still tridiagonal.

It may also be desirable to make the matrix $\mathbf{A}$ symmetric because a symmetric matrix solver uses less CPU time. This can be done by replacing the boundary condition (3.5.3) by the difference form

$$\frac{u_{n+1} - u_n}{\Delta x} + \beta u_n = \gamma. \tag{3.5.13}$$

When u_{n+1} is eliminated between this equation and the last of the set (3.5.6), the result does not alter the coefficient of u_{n-1}:

$$-\frac{1}{\Delta x^2} u_{n-1} + \frac{(1 + \beta \Delta x)}{\Delta x^2} u_n + q u_n = f_n \tag{3.5.14}$$

and symmetry is preserved.

An integrated finite difference method

If the boundary value problem is represented by the differential equation

$$-\frac{d}{dx}\left[p(x) \frac{du}{dx} \right] + q(x)u = f(x) \tag{3.5.15}$$

the following method will produce a symmetric matrix.
Integrate eqn (3.5.15) from $x_{j-1/2}$ to $x_{j+1/2}$ which gives

$$-\left[p(x) \frac{du}{dx} \right]_{x_{j-1/2}}^{x_{j+1/2}} + \int_{x_{j-1/2}}^{x_{j+1/2}} q(x)u \, dx = \int_{x_{j-1/2}}^{x_{j+1/2}} f(x) \, dx$$

and replace this by

$$-p_{j+1/2} \frac{u_{j+1} - u_j}{\Delta x} + p_{j-1/2} \frac{u_j - u_{j-1}}{\Delta x} + \Delta x q_j u_j = \Delta x f_j. \tag{3.5.16}$$

The variable parameter $p(x)$ is thus sampled at intermediate points. Writing this as

$$-a_j u_{j-1} + b_j u_j - c_j u_{j+1} = \Delta x^2 f_j \tag{3.5.17}$$

gives $c_j = p_{j+1/2} = a_{j+1}$ as required for symmetry.
The truncation error of (3.5.16) is still $O(\Delta x^2)$ (when checking this it is necessary first to divide through by Δx so that this difference equation represents the direct replacement of the differential equation).
The set of equations from the finite difference scheme (3.5.6) can be written as

$$\mathbf{Au} - \mathbf{f} = 0 \tag{3.5.18}$$

and the corresponding equation from (3.5.7) as

$$\mathbf{A\hat{u}} - \mathbf{f} = \tau \tag{3.5.19}$$

where $\hat{\mathbf{u}}$, $\mathbf{f}$, τ are the vectors of values of the exact solution, the function $f(x)$, and the truncation error at the grid points, respectively. Then, subtracting and putting the error

$$\mathbf{e} = \hat{\mathbf{u}} - \mathbf{u} \qquad (3.5.20)$$

gives

$$\mathbf{A}\mathbf{e} = \tau \qquad (3.5.21)$$

i.e.

$$\mathbf{e} = \mathbf{A}^{-1}\tau \quad \text{and} \quad \|\mathbf{e}\| \le \|\mathbf{A}^{-1}\|\,\|\tau\| \qquad (3.5.22)$$

where $\|\mathbf{e}\|$, $\|\mathbf{B}\|$ denote the measure of a vector $\mathbf{e}$ or a matrix $\mathbf{B}$ called a norm. See Conte and de Boor (1980) for details of matrix and vector norms. From (3.5.22) because the truncation error τ tends to zero as Δx tends to zero there is convergence provided the matrix $\mathbf{A}$ is invertible. The matrix $\mathbf{A}$ is not invertible, i.e. it is singular, if the boundary conditions for the problem represented by eqn (3.5.1) are $du/dx = 0$ at both ends of the region. This effect can be illustrated by taking $q = 0$ in eqn (3.5.1) and $n = 3$. Using (3.5.8) gives $u_{-1} = u_1$ and $u_3 = u_1$ and the matrix is given by

$$\mathbf{A} = \frac{1}{\Delta x^2} \begin{bmatrix} 2 & -2 & 0 \\ -1 & 2 & -1 \\ 0 & -2 & 2 \end{bmatrix}.$$

Using (3.5.13) gives $u_{-1} = u_0$ and $u_3 = u_2$ and the matrix is given by

$$\mathbf{A} = \frac{1}{\Delta x^2} \begin{bmatrix} 1 & -1 & 0 \\ -1 & 2 & -1 \\ 0 & -1 & 1 \end{bmatrix}.$$

In either case the matrix is clearly singular (adding rows across gives zeros).

All the equations dealt with in this section have been linear. The non-linear problems to be met in later chapters are solved by Picard iteration or by Newton's method (see Appendix 2).

3.6 Partial differential equations—elliptic boundary value problems

A typical example of an elliptic boundary value problem in two dimensions is Poisson's equation:

$$-\frac{\partial^2 h}{\partial x^2} - \frac{\partial^2 h}{\partial y^2} = f(x, y) \qquad (3.6.1)$$

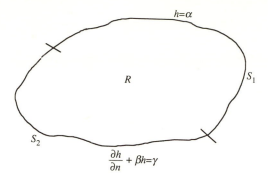

Figure 3.6.1: Region R for a boundary value problem.

which holds in a region R (Fig. 3.6.1) with information given all round the boundary S as follows:

either

$$h = \alpha \text{ is known} \quad \text{(Dirichlet)} \tag{3.6.2}$$

or

$$\frac{\partial h}{\partial n} = \gamma, \quad \text{a 'known-flow' condition} \quad \text{(Neumann)} \tag{3.6.3}$$

or

$$\frac{\partial h}{\partial n} + \beta h = \gamma, \quad \beta \geq 0 \quad \text{(mixed boundary condition).} \tag{3.6.4}$$

Here $\partial h/\partial n$ is the gradient of h in the direction of the outward normal to the boundary S of R, i.e.

$$\frac{\partial h}{\partial n} \equiv \frac{\partial h}{\partial x} n_x + \frac{\partial h}{\partial y} n_y \tag{3.6.5}$$

where $(n_x, n_y)^{\mathrm{T}}$ is the unit outward normal. This is a simplified version of the two dimensions in plan vertically integrated equation for groundwater flow as shown in Appendix 1. The boundary conditions for aquifer problems are usually 'known head' (3.6.2) or 'known flow' (3.6.3).

The reason for writing eqn (3.6.1) with the negative sign on the left-hand side (as in eqn (3.5.1)) is that the resultant matrix of the finite difference equations is then positive definite. Various results connected with the solutions of the equations then follow more easily—see Conte and de Boor (1980) and Varga (1962) for further details.

Suppose the region R is a rectangle covered by a grid formed by lines parallel to the sides of the rectangle spaced Δx, Δy apart. Then a

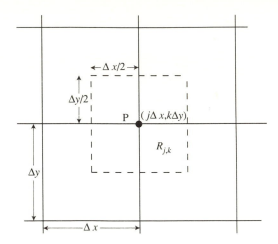

Figure 3.6.2: The integration method—area R_{jk}.

finite difference approximation of the second derivatives in eqn (3.6.1) as in section 3.5 results in a five-point formula centred at $(j\Delta x, k\Delta y)$ given by

$$L_{\Delta x,\Delta y}(h_{j,k}) = \frac{1}{\Delta x^2}[-h_{j-1,k} + 2h_{j,k} - h_{j+1,k}]$$

$$+ \frac{1}{\Delta y^2}[-h_{j,k-1} + 2h_{j,k} - h_{j,k+1}] - f(j\Delta x, k\Delta y) = 0. \quad (3.6.6)$$

The local truncation error can be found as before by applying the difference operator $L_{\Delta x,\Delta y}$ to the exact solution $h(x, y)$ and using the Taylor series expansion in two variables (Conte and de Boor, 1980):

$$h(x + \Delta x, y + \Delta y) = h(x, y) + \Delta x \frac{\partial h}{\partial x} + \Delta y \frac{\partial h}{\partial y}$$

$$+ \frac{\Delta x^2}{2}\frac{\partial^2 h}{\partial x^2} + \Delta x \Delta y \frac{\partial^2 h}{\partial x \, \partial y} + \frac{\Delta y^2}{2}\frac{\partial^2 h}{\partial y^2} + \cdots. \quad (3.6.7)$$

As expected from the corresponding formula for the one-dimensional boundary value problem in section 3.5 the truncation error here is

$$\tau = -\frac{\Delta x^2}{12}\frac{\partial^4 h}{\partial x^4} - \frac{\Delta y^2}{12}\frac{\partial^4 h}{\partial y^4} \quad (3.6.8)$$

i.e. τ is $O(\Delta x^2) + O(\Delta y^2)$ and the finite difference approximation is consistent. Derivative boundary conditions are dealt with by using fictitious nodes and eliminating as in section 3.5.

The integration method in two dimensions

A more general boundary value problem is given by

$$-\frac{\partial}{\partial x}\left[p\,\frac{\partial h}{\partial x}\right] - \frac{\partial}{\partial y}\left[q\,\frac{\partial h}{\partial y}\right] + gh = f(x,y) \qquad (3.6.9)$$

where p, q are variable physical properties, both positive (these could be the components of the transmissivity, see Appendix 1), $g \geq 0$ in a region R, and with boundary conditions on S as before. This can be discretized in a way similar to that used for (3.5.15). Associate with each mesh point P, $(j\Delta x, k\Delta y)$ (Fig. 3.6.2), a region $R_{j,k}$ defined as lying within R and bounded by $x = j\Delta x \pm \frac{1}{2}\Delta x$, $y = k\Delta y \pm \frac{1}{2}\Delta y$. Equation (3.6.9) is now integrated over the region $R_{j,k}$. On the left-hand side of (3.6.9) the differential operator is $-\mathbf{V}\cdot\mathbf{v}$ where

$$\mathbf{v} = \left(p\,\frac{\partial h}{\partial x}, q\,\frac{\partial h}{\partial y}\right)^{\mathrm{T}}$$

The Gauss divergence theorem (Spencer *et al.*, 1977)

$$\int_{R_{j,k}} \mathbf{V}\cdot\mathbf{v}\,\mathrm{d}R = \int_{S_{j,k}} \mathbf{v}\cdot\mathbf{n}\,\mathrm{d}S \qquad (3.6.10)$$

where $S_{j,k}$ is the boundary of $R_{j,k}$ and $\mathbf{n}$ is the outward normal, is used to replace this by

$$\frac{\Delta y}{\Delta x}(h_{j+1,k} - h_{j,k})p_{j+1/2,k} + \frac{\Delta x}{\Delta y}(h_{j,k+1} - h_{j,k})q_{j,k+1/2}$$

$$+ \frac{\Delta y}{\Delta x}(-h_{j,k} + h_{j-1,k})p_{j-1/2,k} + \frac{\Delta x}{\Delta y}(h_{j,k-1} - h_{j,k})q_{j,k-1/2}$$

going anticlockwise round $S_{j,k}$.

This gives a symmetric matrix because of the way the coefficients p and q are evaluated at the midway points. The other terms in (3.6.9) are integrated by taking their values at $(j\Delta x, k\Delta y)$ multiplied by the area of $R_{j,k}$ which is $\Delta x\Delta y$. Thus corresponding to each regular mesh point in R there is an equation of the form

$$D_{j,k}h_{j,k} - W_{j,k}h_{j-1,k} - E_{j,k}h_{j+1,k} - N_{j,k}h_{j,k+1} - S_{j,k}h_{j,k-1} = f_{j,k} \quad (3.6.11)$$

where

$$W_{j,k} = \frac{1}{\Delta x^2}p_{j-1/2,k}, \qquad E_{j,k} = \frac{1}{\Delta x^2}p_{j+1/2,k}$$

$$N_{j,k} = \frac{1}{\Delta y^2}q_{j,k+1/2}, \qquad S_{j,k} = \frac{1}{\Delta y^2}q_{j,k-1/2},$$

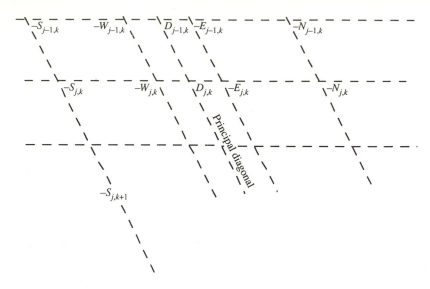

Figure 3.6.3: Structure of the matrix from the integration method.

and

$$D_{j,k} = W_{j,k} + E_{j,k} + N_{j,k} + S_{j,k} + g_{j,k}.$$

Figure 3.6.3 shows the structure of the matrix if the nodes are numbered along the rows of the mesh from left to right and from the bottom row up. The matrix is symmetric because

$$W_{j,k} = E_{j-1,k}, \quad N_{j,k} = S_{j,k+1}, \quad \text{etc.}$$

The truncation error is $O(\Delta x^2) + O(\Delta y^2)$ as before. As with the one-dimensional problem in section 3.5, convergence is safe provided the matrix of the finite difference equations is non-singular. The matrix will be singular if there is only a no-flow Neumann condition all round the boundary as with the problem in section 3.5. Provided there are Dirichlet conditions on part of the boundary the matrix will be positive definite and hence invertible.

The business of dealing with the solution of eqn (3.6.1) or (3.6.9) by finite differences on regions other than an assembly of squares or rectangles is very elaborate. The interested reader can find details in Varga (1962). But regions with curved boundaries are far more easily dealt with by finite elements as described in Chapter 6.

3.7 Initial value problems—parabolic equations

As explained in Chapter 1 a typical parabolic equation in one space dimension is

$$\frac{\partial u}{\partial t} = \sigma \frac{\partial^2 u}{\partial x^2} \tag{3.7.1}$$

where the time $t > 0$, $0 \leq x \leq 1$, $\sigma > 0$.

Here the function $u(x, t)$ is a function of the two independent variables x, t where x is restricted to a finite region and the time t is unrestricted. It is necessary to have information on an initial state $u(x, 0)$ and two boundary conditions on the x variation at $x = 0$ and $x = 1$ of the same type as the boundary conditions listed in (3.5.2)–(3.5.4). This can be called an initial boundary value problem.

Similarly a parabolic equation in two space dimensions

$$\frac{\partial u}{\partial t} = k_x \frac{\partial^2 u}{\partial x^2} + k_y \frac{\partial^2 u}{\partial y^2} \tag{3.7.2}$$

$t > 0$, $k_x, k_y > 0$, where (x, y) is any point in a region R, has to have information all round the boundary of R as well as the initial value $u(x, y, 0)$. The function $u(x, t)$ in (3.7.1) or $u(x, y, t)$ in (3.7.2) could represent the concentration of a diffusing effluent or the height of the water table in an aquifer problem.

The following discussion is concentrated for simplicity on eqn (3.7.1) but the general conclusions apply also to eqn (3.7.2) and the corresponding parabolic equation in three space dimensions. The standard method of solution of eqn (3.7.1) is by 'separation of the variables' (Spencer *et al.*, 1977). Make the substitution

$$u = a_m(t) \exp(imx), \quad i = \sqrt{-1} \tag{3.7.3}$$

to obtain the response to a Fourier mode, mode number m, and suppose the parameter σ is constant. Then the result is

$$\frac{\mathrm{d}a_m}{\mathrm{d}t} = -\sigma m^2 a_m$$

and hence

$$a_m(t) = a_m(0) \exp(-\sigma m^2 t) \tag{3.7.4}$$

(certainly a decaying solution because $\sigma > 0$).

Since eqn (3.7.1) is linear these solutions (3.7.4) can be added to give the general solution

$$u = \sum_{m=-\infty}^{\infty} a_m(0) \exp(-\sigma m^2 t + imx) \tag{3.7.5}$$

where the Fourier series expansion of the initial state $u(x, 0)$ is given by

$$u(x, 0) = \sum_{m=-\infty}^{\infty} a_m(0) \exp(imx)$$

(see, for example, Conte and de Boor, 1980 and Spencer *et al.*, 1977).

   This illustrates the fact that an equation may have an exact solution which is not convenient to compute, being an infinite series. Also this method of breaking the solution down into components can be used with the finite difference scheme which follows.

   The notation u_j^n is used to represent the finite difference approximation to $u(j\Delta x, n\Delta t)$ and a useful shorthand notation is

$$\delta_x^2 u_j^n \equiv u_{j-1}^n - 2u_j^n + u_{j+1}^n. \tag{3.7.6}$$

The corresponding θ method (as in section 3.3) for eqn (3.7.1) then is

$$\frac{u_j^{n+1} - u_j^n}{\Delta t} = \frac{\sigma}{\Delta x^2} [\theta \delta_x^2 u_j^{n+1} + (1 - \theta)\delta_x^2 u_j^n]. \tag{3.7.7}$$

   A widely used method for the analysis of the stability of a finite difference scheme such as (3.7.7) is called von Neumann stability analysis. (This was proposed by von Neumann but not published by him.) The basic idea is like that used in the separation of the variables solution of (3.7.1) described above. That approach supposed that any initial state $u(x, 0)$ could be expressed as an infinite Fourier series expansion. The von Neumann stability analysis supposes that the mesh point values on the initial line, $u(j\Delta x, 0)$, can be expressed as a finite Fourier series:

$$u(j\Delta x, 0) = \sum_{m=0}^{M} a_m \exp(ijm\Delta x)$$

where M is the number of mesh points on the initial line. This leads to trying the substitution into (3.7.7) of a general solution of the form

$$u_j^n = \zeta_m^n \exp(imj\Delta x) \tag{3.7.8}$$

which is of the same pattern as the terms of the exact solution of the differential equation in (3.7.3); m is again the mode number. Substituting into (3.7.6) gives

$$\delta_x^2 \exp(ijm\Delta x) = [\exp(im\Delta x) - 2 + \exp(-im\Delta x)] \exp(imj\Delta x)$$

$$= 2(\cos m\Delta x - 1) \exp(imj\Delta x)$$

i.e.

$$\delta_x^2 \exp(imj\Delta x) = -4s^2 \exp(imj\Delta x) \tag{3.7.9}$$

where

$$s = \sin\left(\frac{m\Delta x}{2}\right). \tag{3.7.10}$$

Thus substituting into (3.7.7) and simplifying gives

$$\xi_m = \frac{1 - 4\mu(1 - \theta)s^2}{1 + 4\mu\theta s^2} \tag{3.7.11}$$

where $\mu = \sigma\Delta t/\Delta x^2$ and ξ_m is the amplification factor of the mth mode.

The numerical solution ought to decay like the exact solution. Hence it is required that $|\xi_m| < 1$, i.e.

$$4\mu(1 - \theta)s^2 < 2 + 4\mu\theta s^2.$$

This leads to the condition

$$4\mu s^2(1 - 2\theta) < 2. \tag{3.7.12}$$

Equation (3.7.12) must be true for all values of $s = \sin(m\Delta x/2)$; hence it must be true when s has its maximum value, unity:

$$2\mu(1 - 2\theta) < 1. \tag{3.7.13}$$

For $\theta \geq 0.5$, (3.7.13) is always satisfied, i.e. there is unconditional stability. For $\theta < 0.5$ there is conditional stability depending on

$$\sigma\frac{\Delta t}{\Delta x^2} = \mu < \frac{1}{2(1 - 2\theta)}. \tag{3.7.14}$$

In particular with the Euler scheme given by $\theta = 0$,

$$\sigma\frac{\Delta t}{\Delta x^2} < \tfrac{1}{2}. \tag{3.7.15}$$

The condition on the time step, (3.7.14), resulting from the von Neumann stability analysis is called the von Neumann stability condition. When using a conditionally stable scheme it is usual to choose Δx first to give the required accuracy in space and then fix Δt to satisfy the stability condition.

The von Neumann stability analysis strictly only applies:

(1) to initial value problems with periodic initial data which can be expanded in a finite Fourier series;

(2) to linear constant coefficient problems such that the Fourier components can be added together;

but it is used very generally. The condition gives a necessary condition for stability for constant coefficient problems regardless of the type of boundary conditions. However, if the coefficients are variable the von Neumann condition can still be applied locally.

The Lax equivalence theorem applies again here. Given a well-posed linear initial value problem and a consistent difference scheme, stability is necessary and sufficient for convergence.

Hence, assuming the original problem is well posed (there is a unique solution and small changes in the data make small changes in the solution), with the difference scheme (3.7.7) and either $\theta \geq 0.5$ or condition (3.7.14) satisfied, reducing the size of Δx and Δt brings the numerical solution nearer to the exact solution.

3.8 The advection–diffusion equation

The equation

$$\frac{\partial C}{\partial t} = D \frac{\partial^2 C}{\partial x^2} - v \frac{\partial C}{\partial x} \tag{3.8.1}$$

is also a parabolic equation. It is typical of an equation representing the one-dimensional advection–diffusion of a contaminant of concentration C; D is the diffusion and v is the velocity. In the non-dimensionalized form the coefficient $\mathbf{R}_e = vL/D$ (supposing $v > 0$), where L is the length of the region, represents the Reynolds number of the flow (see Chapter 10). The second space derivative is approximated by the usual central difference (3.5.5). With a forward difference on the time step and a central difference for the convection term the finite difference equation is

$$C_j^{n+1} = C_j^n + \frac{\Delta tD}{\Delta x^2} [C_{j+1}^n - 2C_j^n + C_{j-1}^n] - \frac{v}{2\Delta x} [C_{j+1}^n - C_{j-1}^n]. \tag{3.8.2}$$

Putting $\Delta C_{j+1/2}^n = C_{j+1}^n - C_j^n$ this can be expressed as

$$C_j^{n+1} = C_j^n + \frac{\Delta tD}{\Delta x^2} \left[1 - \frac{v\Delta x}{2D}\right] \Delta C_{j+1/2}^n - \frac{\Delta tD}{\Delta x^2} \left[1 + \frac{v\Delta x}{2D}\right] \Delta C_{j-1/2}^n. \tag{3.8.3}$$

Put

$$A = \frac{\Delta tD}{\Delta x^2} \left[1 - \frac{v\Delta x}{2D}\right] \quad \text{and} \quad B = \frac{\Delta tD}{\Delta x^2} \left[1 + \frac{v\Delta x}{2D}\right]$$

then (3.8.3) gives

$$C_j^{n+1} = C_j^n + A\Delta C_{j+1/2} - B\Delta C_{j-1/2} \tag{3.8.4}$$

also

$$C_{j+1}^{n+1} = C_{j+1}^n + A\Delta C_{j+3/2}^n - B\Delta C_{j+1/2}^n. \tag{3.8.5}$$

Subtracting (3.8.4) from (3.8.5) gives

$$\Delta C_{j+1/2}^{n+1} = A\Delta C_{j+3/2}^n + (1 - A - B)\Delta C_{j+1/2}^n + B\Delta C_{j-1/2}^n. \tag{3.8.6}$$

Thus, provided A, B, and $(1 - A - B)$ are all greater than or equal to zero,

$$|\Delta C_{j+1/2}^{n+1}| \leq A|\Delta C_{j+3/2}^n| + (1 - A - B)|\Delta C_{j+1/2}^n| + B|\Delta C_{j-1/2}^n|$$

$$\leq (A + 1 - A - B + B)\max_j|\Delta C_{j+1/2}^n| = \max_j|\Delta C_{j+1/2}^n|. \tag{3.8.7}$$

Since this is true for all $|\Delta C_{j+1/2}^{n+1}|$, it must be true for the maximum value of this modulus, i.e.

$$\max_j|\Delta C_{j+1/2}^{n+1}| \leq \max_j|\Delta C_{j+1/2}^n|. \tag{3.8.8}$$

Thus the maximum jump between consecutive values in space is non-increasing—hence any waviness tends to be smoothed out provided A, B, and $(1 - A - B)$ are all non-negative. With $v > 0$ this requires, for $A \geq 0$, the mesh Peclet number

$$\mathbf{P}_e = \frac{v\Delta x}{D} \leq 2 \tag{3.8.9}$$

and for $(1 - A - B) \geq 0$

$$\frac{\Delta t D}{\Delta x^2} \leq \tfrac{1}{2}. \tag{3.8.10}$$

These two conditions can be combined as

$$\frac{v\Delta t}{\Delta x} \leq \frac{2D\Delta t}{\Delta x^2} \leq 1. \tag{3.8.11}$$

Taking the extremes in (3.8.11) gives the condition that the Courant number, $v\Delta t/\Delta x$, is less than or equal to unity (see section 3.10 for further discussion of this). The stability condition (3.8.10) is the same as if the convection term were not there. If the Peclet number condition (3.8.9) is not satisfied then oscillations may persist. In problems where $\mathbf{R}_e$ is large (advection-dominated flows) it can be difficult to satisfy (3.8.9) and 'upstream differencing' can be used. If eqn (3.8.1) is approximated using an upstream difference for the $\partial C/\partial x$ term and a forward difference for the time step the result is

$$C_j^{n+1} = C_j^n + \frac{D\Delta t}{\Delta x^2}[\Delta C_{j+1/2}^n - \Delta C_{j-1/2}^n] - \frac{v\Delta t}{\Delta x}\Delta C_{j-1/2}^n \quad (3.8.12)$$

i.e.

$$C_j^{n+1} = C_j^n + \frac{D\Delta t}{\Delta x^2}\Delta C_{j+1/2}^n - \frac{D\Delta t}{\Delta x^2}\left[1 + \frac{v\Delta x}{D}\right]\Delta C_{j-1/2}^n$$

and

$$C_{j+1}^{n+1} = C_{j+1}^n + \frac{D\Delta t}{\Delta x^2}\Delta C_{j+3/2}^n - \frac{D\Delta t}{\Delta x^2}\left[1 + \frac{v\Delta x}{D}\right]\Delta C_{j+1/2}^n.$$

Subtracting gives

$$\Delta C_{j+1/2}^{n+1} = \frac{D\Delta t}{\Delta x^2}\Delta C_{j+3/2}^n + \left[1 - \frac{D\Delta t}{\Delta x^2}\left(2 + \frac{v\Delta x}{D}\right)\right]\Delta C_{j+1/2}^n$$

$$+ \frac{D\Delta t}{\Delta x^2}\left[1 + \frac{v\Delta x}{D}\right]\Delta C_{j-1/2}^n. \quad (3.8.13)$$

With a similar argument to that used with the central difference the result is

$$\max_j|\Delta C_{j+1/2}^{n+1}| \leq \max_j|\Delta C_{j+1/2}^n|$$

provided

$$1 - \frac{D\Delta t}{\Delta x^2}\left(2 + \frac{v\Delta x}{D}\right) \geq 0$$

i.e.

$$\frac{D\Delta t}{\Delta x^2} \leq \frac{1}{(2 + v\Delta x/D)} \leq \tfrac{1}{2}. \quad (3.8.14)$$

Thus the stability time step limit is reduced but there is no Peclet number effect with the upstream differencing. Note that the first-derivative term must be differenced upstream so if the sign of v changes the difference is in the other direction.

The Peclet number can also have an effect on the finite difference form of the steady-state mass-transport equation

$$D\frac{d^2C}{dx^2} = v\frac{dC}{dx} \quad (3.8.15)$$

with boundary conditions $C = C_0$ at $x = 0$, $C = C_1$ at $x = 1$, say, and D, v constant. The solution for C must have its maximum and minimum values at the boundary, i.e. if $C_1 > C_0$ the solution must increase steadily across the mesh and is then said to be monotonic. This is easily proved as follows.

If there is a local maximum (or minimum) then the first derivative is zero and the second derivative is negative (or positive); but this contradicts eqn (3.8.15) and so cannot be true. With the central difference form eqn (3.8.15) is replaced by

$$(a + b)C_j = aC_{j-1} + bC_{j+1} \qquad (3.8.16)$$

where

$$a = \left[1 + \frac{v\Delta x}{2D}\right], \qquad b = \left[1 - \frac{v\Delta x}{2D}\right].$$

Provided a and b are positive, i.e. the Peclet number is less than two, there will be no waviness in the situation. This can be proved by saying that if $C_j > C_{j-1}$, then from (3.8.16)

$$bC_{j+1} = (a + b)C_j - aC_{j-1} > (a + b)C_j - aC_j = bC_j.$$

Hence $C_{j+1} > C_j$ and the solution continues to increase across the mesh. This means that it cannot have a local maximum and similarly it cannot have a local minimum; the maximum is at one end and the minimum at the other. The solution is then monotonic as it should be and there is no oscillation. There is an example of this effect in connection with salt water intrusion in section 9.5.

The Reynolds number $\mathbf{R_e} = vL/D$ comes from the physics of the problem. The Peclet number $\mathbf{P_e} = v\Delta x/D$ looks like a mesh Reynolds number but its influence depends on the numerical method. As illustrated above the central difference in (3.8.2) is liable to produce oscillation unless $\mathbf{P_e} \leq 2$ but $\mathbf{P_e}$ has no influence on the time step limit in (3.8.10). With the upstream difference in (3.8.12) the influence of the Peclet number is to decrease the time step limit as can be seen in (3.8.14). Note that the relevant mesh Peclet number limit depends on the numerical scheme. Jensen and Finlayson (1980) give Peclet number limits for some other schemes.

If the non-dimensionalized form of eqn (3.8.1) with $\mathbf{R_e} = vL/D$ is approximated with upstream differencing for the advection term and a forward difference for the time step the difference operator is

$$L^{up}_{\Delta x, \Delta t}(C^n_j) = \frac{C^{n+1}_j - C^n_j}{\Delta t} - \frac{\delta^2_x C^n_j}{\Delta x^2} + \mathbf{R_e} \frac{C^n_j - C^n_{j-1}}{\Delta x} = 0. \qquad (3.8.17)$$

Thus the truncation error is

$$\tau_{up} = L^{up}_{\Delta x, \Delta t}(C(j\Delta x, n\Delta t)) = \frac{\Delta t}{2} \frac{\partial^2 C}{\partial t^2} - \mathbf{R_e} \frac{\Delta x}{2} \frac{\partial^2 C}{\partial x^2} + O(\Delta t^2) + O(\Delta x^2). \qquad (3.8.18)$$

Manipulating the differential equation (3.8.1) gives

$$\frac{\partial^2 C}{\partial t^2} = \mathbf{R}_e^2 \frac{\partial^2 C}{\partial x^2} - 2\mathbf{R}_e \frac{\partial^3 C}{\partial x^3} + \frac{\partial^4 C}{\partial x^4}.$$

Hence

$$\tau_{\text{up}} = \frac{\Delta t}{2}\left(\mathbf{R}_e^2 - \mathbf{R}_e \frac{\Delta x}{\Delta t}\right)\frac{\partial^2 C}{\partial x^2} + \frac{\Delta t}{2}\left[-2\mathbf{R}_e \frac{\partial^3 C}{\partial x^3} + \frac{\partial^4 C}{\partial x^4}\right] + O(\Delta t^2) + O(\Delta x^2).$$

$$(3.8.19)$$

However, if eqn (3.8.1) is approximated using the more accurate central difference the truncation error is

$$\tau_{\text{cen}} = \frac{\Delta t}{2}\mathbf{R}_e^2 \frac{\partial^2 C}{\partial x^2}$$

plus the other terms as in (3.8.19). Thus using upstream differencing has the effect of replacing the Reynolds number $\mathbf{R}_e$ by

$$\mathbf{R}_e' = \mathbf{R}_e[(1 - \Delta x/(\mathbf{R}_e \Delta t)]^{1/2} \qquad (3.8.20)$$

and hence the effective Reynolds number is reduced. This produces an artificial smoothing effect on the solution.

Leonard (1979) proposes an algorithm called QUICK (Quadratic Upwind Interpolation for Convective Kinematics). In this the first space derivative is approximated by the central difference (3.8.2) with a one-sided augmentation:

$$\frac{1}{\Delta x}[\tfrac{1}{2}(C_{j+1} - C_{j-1}) + \tfrac{1}{8}(C_{j-2} - 3C_{j-1} + 3C_j - C_{j+1})]. \qquad (3.8.21)$$

Expanding by Taylor series about the values at $j\Delta x$, this expression is equal to

$$\frac{\partial C}{\partial x} + \frac{\Delta x^2}{24}\frac{\partial^3 C}{\partial x^3} + \text{higher-order terms} \qquad (3.8.22)$$

and hence the approximation (3.8.21) has the same order error as the central difference (3.5.5) replacement for the second derivative. Leonard (1979) gives examples to show that this very much reduces the oscillation but avoids excessive smoothing. Of course with an explicit time-stepping scheme there is a stability limit on the time step.

3.9 Initial value problems—hyperbolic equations

The typical hyperbolic equation considered in this section is the kinematic wave equation derived in Chapter 1.

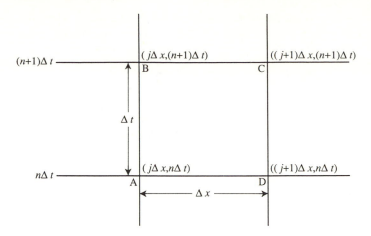

Figure 3.9.1: The box scheme.

A popular finite difference scheme for the numerical solution of this equation is known as the box scheme because it uses the four approximate values of the function at the four points A, B, C, D arranged in a box as in Fig. 3.9.1. The coordinates of these points are $(j\Delta x, n\Delta t)$, $(j\Delta x, (n + 1)\Delta t)$, $((j + 1)\Delta x, (n + 1)\Delta t)$, $((j + 1)\Delta x, n\Delta t)$, respectively. This scheme is also referred to as Preissmann's method (Preissmann, 1961).

The box scheme takes a regular finite difference grid with increments Δx down the slope and Δt in time as in Fig. 3.9.1, with the notation H_j^n for the approximation to $H(j\Delta x, n\Delta t)$. The box scheme is illustrated first on the single kinematic wave equation for overland flow:

$$\frac{\partial H}{\partial t} + c\frac{\partial H}{\partial x} = r \tag{3.9.1}$$

where H is the depth of the water, x is the distance down the slope, t is time, c is the wave speed, and r is the rate of rainfall (as millimetres per hour, for example). The angle of the slope is supposed to be so small that it makes negligible difference whether the depth is measured vertically or perpendicularly to the surface of the slope. The wave speed c is taken to be constant so that the equation (3.9.1) is linear.

First the derivative $\partial H/\partial t$ is expressed as a weighted average of the values obtained using the points AB and the points DC:

$$\frac{\partial H}{\partial t} \approx \frac{(1 - w)}{\Delta t}(H_j^{n+1} - H_j^n) + \frac{w}{\Delta t}(H_{j+1}^{n+1} - H_{j+1}^n) \tag{3.9.2}$$

where w is a parameter taken initially to be such that $0 \le w \le 1$. Substituting the Taylor series expansions in the right-hand side of (3.9.2) shows that the error in this approximation is

$$\frac{\Delta t}{2} \frac{\partial^2 H}{\partial t^2} + w \Delta x \frac{\partial^2 H}{\partial x \, \partial t} + \text{terms in } \Delta x^2 \text{ and } \Delta t^2$$

Similarly the derivative $\partial H/\partial x$ is approximated as a weighted average of the values obtained using the points AD and the points BC:

$$\frac{\partial H}{\partial x} \approx \frac{(1-\theta)}{\Delta x} (H_{j+1}^n - H_j^n) + \frac{\theta}{\Delta x} (H_{j+1}^{n+1} - H_j^{n+1}) \qquad (3.9.3)$$

where θ is a parameter, $0 \le \theta \le 1$. The error in the approximation (3.9.3) is

$$\frac{\Delta x}{2} \frac{\partial^2 H}{\partial x^2} + \theta \Delta t \frac{\partial^2 H}{\partial x \, \partial t} + \text{terms in } \Delta x^2 \text{ and } \Delta t^2. \qquad (3.9.4)$$

Thus in the general box scheme eqn (3.9.1) is approximated by

$$\frac{(1-w)}{\Delta t} (H_j^{n+1} - H_j^n) + \frac{w}{\Delta t} (H_{j+1}^{n+1} - H_{j+1}^n)$$

$$+ c \left(\frac{(1-\theta)}{\Delta x} (H_{j+1}^n - H_j^n) + \frac{\theta}{\Delta x} (H_{j+1}^{n+1} - H_j^{n+1}) \right) - r = 0. \quad (3.9.5)$$

It is assumed that:

(1) the initial condition $H(x, 0)$ is known so that $H_j^0 = H(j\Delta x, 0)$ for all values of j; and

(2) $H(0, t)$ is known at one end of the region so that $H_0^n = H(0, n\Delta t)$.

Then the box scheme (3.9.5) can be used to step out the solution from one end to the other for successive time steps.

Equation (3.9.5) has an error due to truncating the Taylor series (Conte and de Boor, 1980), given by

$$\tau = \frac{\Delta t}{2} \left(\frac{\partial^2 H}{\partial t^2} + 2c\theta \frac{\partial^2 H}{\partial x \, \partial t} \right) + \frac{\Delta x}{2} \left(c \frac{\partial^2 H}{\partial x^2} + 2w \frac{\partial^2 H}{\partial x \, \partial t} \right) + \text{terms in } \Delta x^2 \text{ and } \Delta t^2.$$

$$(3.9.6)$$

But with constant c and r eqn (3.9.1) can be differentiated partially with respect to t and x in turn to give

$$\frac{\partial^2 H}{\partial t^2} = -c \frac{\partial^2 H}{\partial x \, \partial t} = c^2 \frac{\partial^2 H}{\partial x^2}. \qquad (3.9.7)$$

Substituting from (3.9.7) into (3.9.6) then gives

$$\tau = \frac{1}{2}\frac{\partial^2 H}{\partial x\, \partial t}[(2\theta - 1)c\Delta t + (2w - 1)\Delta x] + \text{terms in } \Delta x^2 \text{ and } \Delta t^2. \quad (3.9.8)$$

This shows that the box scheme is most accurate when $\theta = w = 0.5$. The effect is then that the expressions for the first derivatives are each centred at the midpoint of a side of a box and the averaging with $\theta = w = 0.5$ centres the whole expression at the midpoint of the box. This consideration is useful in dealing with the non-linear case when the velocity c depends on the solution. Several variations of the box scheme for the non-linear case are given in Chapter 4. The box scheme is consistent because the truncation error (3.2.26) or (3.2.8) tends to zero as the time and space steps, Δt and Δx, tend to zero.

The box scheme can similarly be applied to the solution of the kinematic wave equation expressed in terms of the discharge q as obtained in section 1.2:

$$\frac{\partial q}{\partial t} + c\left(\frac{\partial q}{\partial x} - br\right) = 0 \quad (3.9.9)$$

using the values of q at the four corners of the box ABCD in Fig. 3.9.1.

3.10 Stability and convergence illustrated on the box scheme

The box finite difference scheme can be tested by the von Neumann stability analysis method as described in section 3.7 by looking at its response to a Fourier-mode-type solution

$$H_j^n = \xi^n \exp(\mathrm{i}jp\Delta x) \quad (3.10.1)$$

where $i = \sqrt{-1}$ and p is the mode number. From (3.10.1) we have

$$H_j^{n+1} = \xi H_j^n$$

and hence the condition for stability is that $|\xi| \leq 1$ for then the solution remains bounded.

It is sufficient to examine the stability on the homogeneous scheme, i.e. eqn (3.9.5) with $r = 0$; substituting from (3.10.1), putting $\lambda = c\Delta t/\Delta x$, and cancelling $\xi^n \exp(\mathrm{i}jp\Delta x)$ gives

$$(\xi - 1)[1 - w + w \exp(\mathrm{i}p\Delta x)] + \lambda[\exp(\mathrm{i}p\Delta x) - 1](1 - \theta + \xi\theta) = 0.$$

Putting $1 - w - \lambda\theta = a$ and $w + \lambda\theta = b$ and rearranging gives

$$\xi(a + b\hat{c} + \mathrm{i}b\hat{s}) = a + \lambda + (b - \lambda)\hat{c} + \mathrm{i}(b - \lambda)\hat{s}$$

where $\hat{c} = \cos p\Delta x$ and $\hat{s} = \sin p\Delta x$. Hence

$$|\xi|^2 = \frac{[a + \lambda + (b - \lambda)\hat{c}]^2 + (b - \lambda)^2\hat{s}^2}{(a + b\hat{c})^2 + b^2\hat{s}^2} \qquad (3.10.2)$$

and the stability condition $|\xi| \leq 1$ requires that

$$[a + \lambda + (b - \lambda)\hat{c}]^2 + (b - \lambda)^2\hat{s}^2 \leq (a + b\hat{c})^2 + b^2\hat{s}^2.$$

Multiplying out and cancelling 2λ which is certainly positive this condition reduces to

$$(a - b + \lambda)(1 - \hat{c}) \leq 0$$

where $(1 - \hat{c}) = (1 - \cos p\Delta x)$ is certainly not negative. Hence this condition is satisfied if

$$a - b + \lambda \leq 0.$$

Substituting for a and b then gives

$$(1 - 2\theta)\lambda + (1 - 2w) \leq 0. \qquad (3.10.3)$$

With the three-point scheme using only the points ACD as in Fig. 3.10.1 with $\theta = 0$ and $w = 1$, condition (3.10.3) gives the limit on the time step

$$\lambda = c\frac{\Delta t}{\Delta x} \leq 1. \qquad (3.10.4)$$

Referring back to the method of characteristics described in Chapter 2 this is the condition that the numerical scheme does not step out beyond the characteristic through the point A. It is also the condition that the three-point scheme remains stable.

With the four-point box scheme we only have to have $\theta \geq 0.5$, $w \geq 0.5$, and condition (3.10.3) is satisfied for any size of time step. The box scheme is particularly easy to use because it is explicit in the sense that with the first-order derivative in space there can only be one

(a)

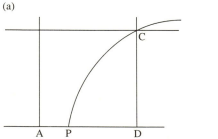

(b)

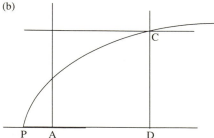

Figure 3.10.1: The CFL condition.

boundary condition. If $H(0, t)$ is known at the top of the slope then H_{j+1}^{n+1} is the only unknown in eqn (3.9.5) where the scheme is marching from left to right in Fig. 3.9.1. Similarly if eqn (3.9.9) is being approximated and the value of q is known at one end of the slope the scheme can be marched along from there.

This leads on to the important concept of the CFL (Courant–Friedrichs–Lewy, 1928, 1967) condition for convergence of a finite difference scheme for the solution of a hyperbolic equation such as the kinematic wave equation (3.9.1). This is the condition for the numerical solution to converge to the exact solution as Δt and Δx are decreased (this is not the same as the convergence of any iteration which is taking place). The discussion of stability above depends on starting at a point A on the initial line and observing whether the numerical method has a limit on the time step in order to ensure that the numerical solution does not blow up. The CFL convergence criterion depends on observing that because this is a hyperbolic equation the solution at a point C, $(x_j, \Delta t)$ (Figure 3.10.1), is governed by the solution on the characteristic through C which has started from some point P on the initial line. If the solution at C is calculated using the grid values at A and D and P lies within AD (Fig. 3.10.1(a)) (i.e. the value at P can be regarded as interpolated between the values at A and D and therefore dependent on them), then this is in accordance with the physics of the problem. But if the solution at C is calculated using the values at A and D and P does not lie between these points (Fig. 3.10.1(b)) then the initial value at P can change (e.g. because of a spatial variation in the rainfall), without altering the numerical solution at C; this contradicts the physical conditions implicit in the theory of characteristics. Since the local slope of the characteristic is $1/c$ (eqn. (1.5.16)), the CFL condition here is

$$\frac{\Delta t}{\Delta x} \leq \frac{1}{c}, \quad \text{i.e. } \mathbf{C}_N = \frac{c\Delta t}{\Delta x} \leq 1. \tag{3.10.5}$$

$\mathbf{C}_N$ is called the Courant number. The condition (3.10.5) is the condition that the domain of dependence of the solution at C lies within the region covered by the three-point scheme, i.e. the characteristic through C starts from a point on AD. The condition (3.10.5) is also the same as the condition (3.10.4) for the three-point scheme to be stable.

The four-point box scheme is using values at B (previously calculated), A, and D in Fig. 3.9.1. to calculate the solution at C and hence the domain of dependence of the scheme certainly includes the characteristic through C in either case a or b. Thus the four-point box scheme always satisfies the CFL condition and illustrates the result that every such stable scheme is convergent.

3.11 The finite volume or cell-differencing method

This is a finite difference method for elliptic or parabolic problems
which starts by dividing the region into subregions which may be
called finite volumes or cells. These are usually quadrilaterals (in two
dimensions) or boxes (in three dimensions) in surface flow (see Burns
and Wilkes, 1987*)* but may be of more general shape (see Narasimhan
and Witherspoon (1976) in connection with the integrated finite
difference method for subsurface flow, section 5.4). All variables are
defined at the centre of a cell. The equations of motion are integrated
over the cell and the Gauss divergence theorem (Spencer *et al.*, 1977)
used as in the integration method (section 3.6) to reduce the space-
differentiated terms to integrals on the boundary of the cell. When the
equation has a simple form as in (3.6.9), the cell is a square as R_{ij} on
Fig. 3.6.2, and the principal directions of the transmissivity are aligned
with the sides of the cell, the boundary integral terms being easily
approximated by finite difference expressions as shown there. But
when these conditions do not hold and/or when there are first
derivatives in the original equation, the result is that some of the
boundary derivatives can have components along the sides of the cell
which leads to more complicated finite difference expressions. This
situation is more easily handled by the finite element method (Chapter
6) because there it is the basis functions which are differentiated. The
finite volume method is referred to in connection with water-quality
modelling in section 9.2.

4 Numerical solution of surface flow equations by finite differences

4.1 Introduction

This chapter discusses finite difference methods applied to the solution of surface flow equations. The examples chosen are those which can be solved by fairly simple methods. They are arranged in the order in which they occur in Table 1.1.1. The first, lateral velocity distribution across a channel, is an example of an ordinary differential equation boundary value problem. The box scheme, introduced in section 3.8, is applied in section 4.2 to the non-linear hyperbolic kinematic wave equation. Section 4.3 includes the box scheme and the Abbott scheme applied to the solution of the St Venant equations. Section 4.4 discusses the limitations on the use of these schemes and section 4.5 lists some alternatives. Section 4.5 describes in outline only some more sophisticated methods for the problem of vertically integrated two-dimensional surface flow.

In section 1.3 the equation for the velocity distribution across the channel in steady surface flow is obtained:

$$-\frac{\mathrm{d}}{\mathrm{d}x}\left(\varepsilon H \frac{\mathrm{d}u}{\mathrm{d}x}\right) + \frac{k}{\rho}|u|u = gHs \qquad (4.1.1)$$

where u is the depth-averaged velocity, ε is the turbulence exchange parameter, H is the depth, ρ is the density, $k|u|u$ is the friction force per unit area, g is the acceleration due to gravity, s is the slope, and the x axis is taken horizontally across the channel which stretches from $x = 0$ to $x = L$ (Fig. 1.3.1). This is an example of an ordinary differential equation boundary value problem. The boundary conditions are $u = 0$ at $x = 0, L$. $H = H(x)$ is supposed known for a particular height of the water surface above datum (stage) and information about the bed of the channel and adjacent flood plain. Using the integration method in section 3.5 the finite difference form of (4.1.1) can be taken as

$$-\varepsilon H_{j+1/2}\frac{(u_{j+1} - u_j)}{\Delta x} + \varepsilon H_{j-1/2}\frac{(u_j - u_{j-1})}{\Delta x} + \frac{\Delta x k}{\rho}|u_j|u_j = gH_j s \Delta x \qquad (4.1.2)$$

The equation (4.1.2) is non-linear and hence an iterative method must be used, for example

$$-\varepsilon H_{j+1/2}\frac{(u_{j+1}^i - u_j^i)}{\Delta x} + \varepsilon H_{j-1/2}\frac{(u_j^i - u_{j-1}^i)}{\Delta x} + \frac{\Delta x k}{\rho}|u_j^{i-1}|u_j^i = gH_j s\Delta x$$

(4.1.3)

where i is the iteration number. Wark *et al.* (1990) use Newton's method (Appendix 2) starting from a 'seed' solution obtained by putting $\varepsilon = 0$ as a first approximation. At each iteration a tridiagonal matrix solver can be used—see for example Conte and de Boor (1980).

4.2 The non-linear kinematic wave equation

As shown in section 1.5 the single kinematic wave equation representing sheet flow is of the form

$$\frac{\partial H}{\partial t} + c(H)\frac{\partial H}{\partial x} = r$$

(4.2.1)

where $c = \beta\alpha H^{\beta-1}$; $\beta = 5/3, 3/2$ for the Manning and Chezy resistance formulae respectively. The parameter α includes the slope and the resistance coefficients (see French (1987) for details). Equation (4.2.1) can be expressed in terms of the discharge q in a form also suitable for channel flow as

$$\frac{\partial q}{\partial t} + c(q)\left(\frac{\partial q}{\partial x} - br\right) = 0$$

(4.2.2)

where the form of $c(q)$ depends on the resistance formula. This can be solved by the box scheme (section 3.9) but a decision has to be made on how to treat the non-linear term. One possibility is to take

$$(1-w)\frac{(q_j^{n+1} - q_j^n)}{\Delta t} + w\frac{(q_{j+1}^{n+1} - q_{j+1}^n)}{\Delta t}$$

$$+ (1-\theta)c(\bar{q}^n)\left[\frac{(q_{j+1}^n - q_j^n)}{\Delta x} - r\right] + \theta c(\bar{q}^{n+1})\left[\frac{(q_{j+1}^{n+1} - q_j^{n+1})}{\Delta x} - r\right] = 0$$

(4.2.3)

where $\bar{q}^n = (q_j^n + q_{j+1}^n)/2$.

This non-linear equation can be solved for q_{j+1}^{n+1} using an iteration method; see Appendix 2. The truncation error of this box scheme is the same as for constant kinematic wave velocity c, i.e. as in eqn (3.8.8) with H replaced by q:

$$\tau = (w - 0.5)\Delta x\frac{\partial^2 q}{\partial x\,\partial t} - (\theta - 0.5)\Delta t\frac{\partial^2 q}{\partial t^2} + \text{second-order terms.} \quad (4.2.4)$$

Applying the von Neumann stability analysis as in section 3.9 for a particular time step we put $q_j^n = \zeta^n \exp(ipj\Delta x)$ and substitute $\lambda_n = c(\bar{q}^n)\Delta t/\Delta x$. The stability condition for $\xi \leq 1$ is then

$$(1 - \theta)\lambda_n - \theta\lambda_{n+1} \leq 2w - 1. \tag{4.2.5}$$

For $w = 1$ and $\theta = 0$, (4.2.3) gives the explicit three-point scheme which is conditionally stable. Equation (4.2.5) gives the condition on the time step as coming from

$$c(\bar{q}^n)\frac{\Delta t}{\Delta x} = \lambda_n \leq 1$$

which is the same as the local CFL condition (section 3.10). No iteration is needed to deal with the non-linearity here but the value of $c(\bar{q}^n)$ has to be taken into account in the time step control. The right-hand side of (4.2.5) is non-negative if $w \geq 0.5$. If, for instance, $w = 1$, the stability condition becomes

$$(1 - \theta)\lambda_n - \theta\lambda_{n+1} \leq 1. \tag{4.2.6}$$

For $\theta \neq 1$ the condition becomes

$$c_n \leq \frac{\theta}{1 - \theta}c_{n+1} + \frac{\Delta x}{\Delta t(1 - \theta)} \tag{4.2.7}$$

where $c_n = c(\bar{q}^n)$. With a large time step and decreasing flow this will probably not be satisfied. For the more accurate version of the box scheme with $w = \theta = 0.5$ the stability condition (4.2.5) requires that

$$c_n \leq c_{n+1} \tag{4.2.8}$$

which again will not be satisfied with decreasing flow. The solution is to take a single averaged value for the argument of c, i.e. to use $\bar{c} = c(\bar{q}^{n+1/2})$ where

$$\bar{q}^{n+1/2} = \tfrac{1}{4}(q_j^n + q_j^{n+1} + q_{j+1}^n + q_{j+1}^{n+1}). \tag{4.2.9}$$

The stability condition (4.2.5) is then

$$(1 - 2\theta)\bar{\lambda} \leq 2w - 1 \tag{4.2.10}$$

and there is now unconditional stability for $w \geq 0.5$, $\theta \geq 0.5$ as in the case of constant kinematic velocity.

Arnold (1989) uses the method (4.2.3) with $w = 1$ without meeting this instability problem but his program is designed to halve the time step if any difficulty is met. Arnold solves eqn (4.2.2) for sheet flow taking $b = 1$ so that q is the discharge per unit width of slope. The truncation error (4.2.4) then becomes

$$\tau = \frac{\Delta x}{2} \frac{\partial^2 q}{\partial x \, \partial t} - (\theta - 0.5)\Delta t \frac{\partial^2 q}{\partial t^2} + \text{second-order terms.} \qquad (4.2.11)$$

The iteration used here is straightforward: an initial estimate of q_{j+1}^{n+1} is used to calculate $c(\bar{q}^{n+1})$, the equation (4.2.3) is solved for the next estimate, $c(\bar{q}^{n+1})$ is adjusted and so on until convergence is reached. In Arnold (1989) the difference between the two last successive iterates and the residual when the last iterate is substituted into eqn (4.2.3) are required to be less than their specified tolerances within 10 iterations. Otherwise the time step is halved and the process repeated. It is interesting to look at the difficulties encountered in the iteration here which are typical of many problems.

1. Calculation of the first space step at the time level when the rain starts: if the iteration is generally started with the initial value of q_{j+1}^{n+1} taken equal to q_{j+1}^{n}, then since before the rain starts there is no flow, the first estimate of $c(\bar{q}^{n+1})$ is zero. Also $q_0^n = 0$ for all n, and hence eqn (4.2.3) gives q_1^{n+1} as zero too. The remedy used by Arnold is to take the initial value of q_{j+1}^{n+1} equal to the rainfall rate r and thus this particular difficulty is removed.

2. Calculation of the first space step when the rain stops: when the rainfall stops eqn (4.2.3) (with $w = 1$) gives for the first space step

$$\frac{q_1^{n+1} - q_1^n}{\Delta t} + (1 - \theta)c(\bar{q}^n)\left[\frac{q_1^n - 0}{\Delta x}\right] + \theta c(\bar{q}^{n+1})\left[\frac{q_1^{n+1} - 0}{\Delta x}\right] = 0. \quad (4.2.12)$$

In the iteration to calculate q_1^{n+1} the first iterate now is $r = 0$, and $c(\bar{q}^{n+1}) = 0$. Thus the value of the discharge at one space step down the slope is given by

$$q_1^{n+1} = q_1^n - \Delta t(1 - \theta)c(\bar{q}^n)\frac{q_1^n}{\Delta x}. \qquad (4.2.13)$$

This gives a negative value when

$$1 - (1 - \theta)c(\bar{q}^n)\frac{\Delta t}{\Delta x} < 0. \qquad (4.2.14)$$

There is no trouble if $\theta = 1$ but otherwise there is a negative value when the time step is large enough to make

$$\Delta t > \frac{\Delta x}{c(\bar{q}^n)(1 - \theta)}. \qquad (4.2.15)$$

The remedy here is an extra condition in the program so that if a negative value of q appears the time step is halved.

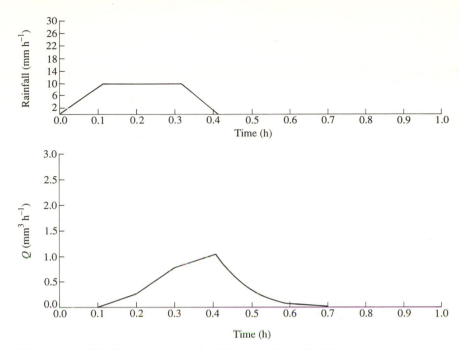

Figure 4.2.1: Discharge at bottom of slope with gradual increase and decrease of rainfall.

Figures 4.2.1–4 show results from Arnold (1989) with a slope $s = 0.1$, length $L = 100$ m, and Chezy roughness with a coefficient appropriate for a car park. Figure 4.2.1 shows the effect on the discharge Q at the foot of the slope of a gradual increase and decrease in rainfall and this can be compared with Fig. 1.5.4. The other figures show results with rainfall at a constant rate of 10 mm per hour which starts at 0.1 hours and ends at 0.4 hours. Figure 4.2.2 shows the error in the magnitude of the peak discharge Q_p against Δt for $\theta = 0.5$, $\Delta x = 0.1$ m. The value of θ, $0.5 \leq \theta \leq 1.0$, has negligible effect on the error in the magnitude of Q_p. Figure 4.2.3 shows the error in the time t_p to the peak discharge against Δt with $\theta = 0.5$, $\Delta x = 0.1$ m. Figure 4.2.4 shows the error in t_p against θ, verifying that $\theta = 0.5$ gives the most accurate value.

4.3 The numerical solution of the St Venant equations

The St Venant equations for continuity of mass and momentum obtained in Chapter 1 can be expressed in different forms with different pairs of dependent variables; for convenience they are written again

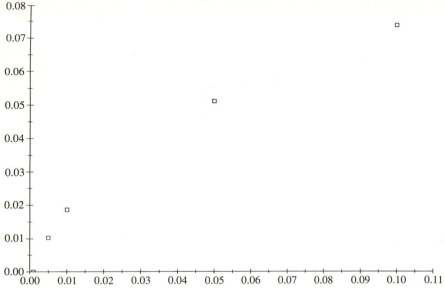

Figure 4.2.2: Error in peak discharge against Δt.

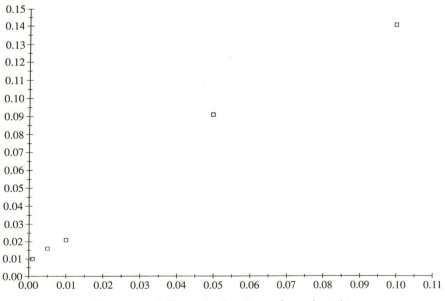

Figure 4.2.3: Error in time to peak against Δt.

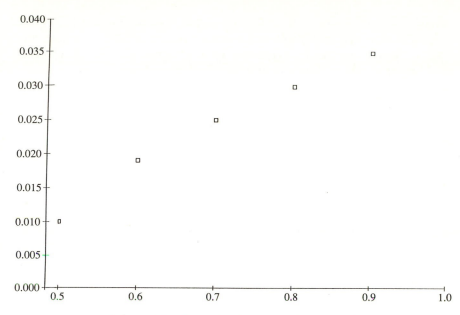

Figure 4.2.4: Error in time to peak against θ.

here. A form suitable for channel flow gives the continuity equation expressed in terms of the cross-sectional area A and the discharge q:

$$\frac{\partial A}{\partial t} + \frac{\partial q}{\partial x} = br \tag{4.3.1}$$

where t is time, x distance down slope, b breadth of the flow and r rate of inflow of water (rainfall, inflow from channel banks, etc.), and the momentum equation as

$$\frac{\partial q}{\partial t} + \frac{\partial}{\partial x}\left(\frac{q^2}{A}\right) + gA\left(\frac{\partial h}{\partial x} + s_f\right) = 0 \tag{4.3.2}$$

where h is the height of the water surface above datum (the stage), $h = H + (L - x)s$, (Fig. 4.3.1), H is the water depth, L the length of

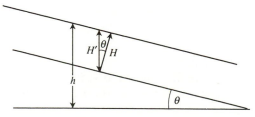

Figure 4.3.1: H–h relationship, $H \approx H'$.

slope, $s = \sin \theta$ where θ is the slope angle (assumed small), s_f is the friction slope, and g is the acceleration due to gravity. Alternatively the St Venant equations expressed in a form suitable for flood plain flow in terms of the depth H and the discharge q are

$$b \frac{\partial H}{\partial t} + \frac{\partial q}{\partial x} = br \qquad (4.3.3)$$

where $b = \partial A/\partial h$, and

$$\frac{\partial q}{\partial t} + \frac{\partial}{\partial x}\left(\frac{q^2}{A}\right) + gA\left(\frac{\partial H}{\partial x} + s_f - s\right) = 0. \qquad (4.3.4)$$

There are many finite difference versions of the St Venant equations depending on the form chosen for the differential equations before the discretization takes place. These versions are the result of combining eqns (4.3.3) and (4.3.4) in different ways.

The friction slope term s_f represents the losses due to friction at the channel bed. Over a certain range of velocities and depths of flow this can be taken as given by the formula (1.2.4) as explained in section 1.2.

For an initial value problem initial conditions must be specified and the solution stepped out from these. The ideal is an initial steady state which is easy to represent as in the simple example of sheet flow down a slope described in section 4.2. It can be much more difficult with an actual river model. The pseudo time stepping advocated by Cunge *et al.* (1980) starts with a rough guess at the initial conditions and runs with a small time step Δt until all the waves generated by the initial noise have disappeared out of the system. The time step Δt is then increased until the change in solution is negligible. Evans and Whitlow (1991) use what they call an 'elaboration' of the fourth-order Runge–Kutta method (section 3.3) to solve the steady-state St Venant equations for the initial conditions and they claim this performs well with Froude numbers as high as 0.98.

It is also necessary to have a boundary condition for each of the unknowns, q and H, specified at an end of the region. The St Venant equations are based on the assumptions of subcritical flow where the Froude number $\mathbf{F} < 1$. As shown in Chapter 2 on the method of characteristics, subcritical flow where the Froude number $\mathbf{F} = u/\sqrt{(gH)} < 1$ corresponds to a situation where there are two families of characteristics, one pointing downstream and the other pointing upstream. Hence the boundary conditions for q (or u) and H can be given at opposite ends of the region as well as their initial values. For instance, the inflow hydrograph at the head of a river might give $q(t)$ and if the sea is the downstream boundary the tidal level $H(t)$ might be given.

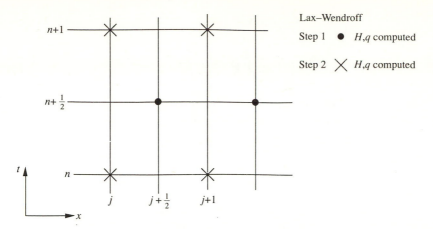

Figure 4.3.2: Lax–Wendroff mesh.

This means that there is a set of simultaneous non-linear equations to be solved for each time step. Having boundary conditions at both ends of the region acts as a control on the solution.

An often-used finite difference method for the solution of the St Venant equations is the Lax–Wendroff explicit scheme using the mesh illustrated in Fig. 4.3.2. Supposing that the values of q_j^n and H_j^n are known, the values of $H_{j+1/2}^{n+1/2}$ and $q_{j+1/2}^{n+1/2}$ are calculated from the St Venant continuity and momentum equations using approximations of the form

$$\frac{\partial f}{\partial t} \approx \frac{2}{\Delta t}\left[f_{j+1/2}^{n+1/2} - \frac{(f_{j+1}^n + f_j^n)}{2}\right] \qquad (4.3.5)$$

for the time derivatives with the space derivatives formed from differences at the n level and linearized averaged values for the q^2/A term, i.e. from (4.3.3) the equation for $H_{j+1/2}^{n+1/2}$ is

$$\tfrac{1}{2}(b_{j+1}^n + b_j^n)[H_{j+1/2}^{n+1/2} - \tfrac{1}{2}(H_{j+1}^n + H_j^n)]\frac{2}{\Delta t} + \frac{1}{\Delta x}(q_{j+1}^n - q_j^n)$$

$$= \tfrac{1}{2}(b_{j+1}^n + b_j^n)\tfrac{1}{2}(r_{j+1/2}^n + r_{j+1/2}^{n+1})$$

and the equation for $q_{j+1/2}^{n+1/2}$ is (4.3.4) in the form

$$\frac{2}{\Delta t}[q_{j+1/2}^{n+1/2} - \tfrac{1}{2}(q_{j+1}^n + q_j^n)] + \frac{1}{\Delta x}\left[\left(\frac{q^2}{A}\right)_{j+1}^n - \left(\frac{q^2}{A}\right)_j^n\right]$$

$$+ \frac{g}{2}(A_{j+1}^n + A_j^n)\left[\frac{1}{\Delta x}(H_{j+1}^n - H_j^n) + \tfrac{1}{2}[(s_f)_{j+1}^n + (s_f)_j^n] - s\right] = 0.$$

Then the St Venant equations are used again now with time derivatives approximated as in

$$\frac{\partial f}{\partial t} \approx \frac{f_j^{n+1} - f_j^n}{\Delta t} \tag{4.3.6}$$

and the space derivatives formed from differences at the $n + 1/2$ level with corresponding averaged values of the q^2/A and s_f terms. The value of H_j^{n+1} is given by

$$\frac{1}{\Delta t} b_j^n (H_j^{n+1} - H_j^n) + \frac{1}{\Delta x} (q_{j+1/2}^{n+1/2} - q_{j-1/2}^{n+1/2}) = \tfrac{1}{2} b_j^n (r_j^{n+1} + r_j^n)$$

(the representation of the inflow r depends on the form of the data available) and the value of q_j^{n+1} is given by

$$\frac{1}{\Delta t}(q_j^{n+1} - q_j^n) + \frac{1}{\Delta x}\left[\left(\frac{q^2}{A}\right)_{j+1/2}^{n+1/2} - \left(\frac{q^2}{A}\right)_{j-1/2}^{n+1/2}\right] + \frac{g}{2}(A_{j+1/2}^{n+1/2} + A_{j-1/2}^{n+1/2})$$

$$\times \left[\frac{1}{\Delta x}(H_{j+1/2}^{n+1/2} - H_{j-1/2}^{n+1/2}) + \tfrac{1}{2}[(s_f)_{j+1/2}^{n+1/2} + (s_f)_{j-1/2}^{n+1/2}] - s\right] = 0.$$

There is only one unknown per equation so that the method is explicit. It is only conditionally stable with the time step limited by

$$\frac{\Delta t}{\Delta x} \leq \frac{1}{v + \sqrt{(gH)}}. \tag{4.3.7}$$

With a staggered mesh there is always a problem of how to deal with the boundary conditions.

Two popular methods for solving the St Venant equations are the box scheme (Preissmann, 1961) and the Abbott scheme (Abbott and Ionescu, 1967). The box scheme uses the box grid as in the solution of the kinematic wave equation in section 4.2, i.e. the dependent variables H (or h) and Q are located at the corners of the box and usually all functions and derivatives are centred in the box. It is thus convenient to keep the dependent variables in the term q^2/A together rather than expand into products including $\partial q/\partial x$ and $\partial A/\partial x$. This idea is used by Evans (1977), Samuels (1985), Samuels and Skeels (1990); Samuels calls this term the convection acceleration term.

In the application of a finite difference scheme it is usual to choose the spatial grid size Δx first. With the box scheme Δx can easily be varied to take account of hydraulic structures, bridges, etc. It is generally supposed that higher values of the Froude number, $u/\sqrt{(gH)}$, require smaller Δx. Samuels (1990) gives a set of guidelines for the choice of cross-sections in river models. Evans and Whitlow (1991) argue that it is better to avoid small Δx because of the correspondingly larger values of the Courant number $u\Delta t/\Delta x$. See Abbott, 1979 for a discussion on the effect of a large Courant number here.

In its general form the box scheme can be applied to eqns (4.3.1) and (4.3.2) or to eqns (4.3.3) and (4.3.4) with the derivatives of any function $f(x, t)$ approximated as

$$\frac{\partial f}{\partial t} \approx \frac{(1-w)}{\Delta t}(f_j^{n+1} - f_j^n) + \frac{w}{\Delta t}(f_{j+1}^{n+1} - f_{j+1}^n)$$

$$\frac{\partial f}{\partial x} \approx \frac{(1-\theta)}{\Delta x}(f_{j+1}^n - f_j^n) + \frac{\theta}{\Delta x}(f_{j+1}^{n+1} - f_j^{n+1})$$

(4.3.8a)

and the function $f(x, t)$ itself approximated as

$$f \approx \tfrac{1}{2}[(1-\theta)(f_{j+1}^n + f_j^n) + \theta(f_{j+1}^{n+1} + f_j^{n+1})]$$ (4.3.8b)

where f is any of the terms that occur in the equations, i.e. in (4.3.1) and (4.3.2) these are A, q, q^2/A, and h as a function of A depending on the shape of the cross-section. In (4.3.3) and (4.3.4) the terms are H, q, q^2/A, and A as a function of H. Equation (4.3.8a) is usually used with $w = 0.5$.

Then application of the box scheme (Fig. 3.8.1) necessitates the solution of non-linear equations for the values of q_j^n, H_j^n. There are four unknowns q_j^{n+1}, H_j^{n+1}, q_{j+1}^{n+1}, H_{j+1}^{n+1} in each equation. Using the Newton–Raphson method (Appendix 2) gives a five-diagonal banded matrix system to solve and there is a fast algorithm for this (see Duff *et al.* (1989) on band methods).

With the box scheme it is simple to arrange the program so that the space and time steps, Δx and Δt, can be varied. The time step can be controlled by limiting the number of iterations allowed in the solution of the non-linear equations. If convergence is not reached then the computation can be repeated with the time step halved.

The optimum value of the time step for lowest error (Price, 1974) is when

$$\frac{\Delta x}{\Delta t} \approx c_c, \quad \text{where } c_c = \max\left(\frac{q}{A} + \sqrt{(gH)}\right)$$ (4.3.9)

i.e. when $\Delta x/\Delta t$ has the same slope as the characteristic (Chapter 2). When the water in the channel overflows the banks there will be different values for the optimum Δt in the channel and in the flood plain because of the different wave speeds. Samuels (1977) recommends treating the flow in the main channel and the flood plains separately with an interchange q_b of water per unit length between the channel and the flood plain given by

$$q_b = C_b A_c (gA_c/B_c)^{1/2} \qquad (4.3.10)$$

where A_c and B_c are the area and surface width of the flow over the bank and the coefficient C_b depends on the water level on either side. Pender (1992) adjusts the time step for the flood plain calculation independently from the main channel solution in order to avoid numerical oscillation. French (1987, p. 572), gives a method of combining the flow with proportions of it in the channel and the left and right flood plains. The solution of the equation for the lateral distribution of velocity across the channel and flood plain in the steady-state case is discussed in section 4.1.

Other references which use the box (Preissmann) scheme are: Cunge (1989) to obtain the velocities for modelling the transport of suspended sediment; Neat *et al.* (1989) for the modelling of flood effects on the rivers Aire and Calder; Goodwin *et al.* (1989) for modelling the flood from a dam break.

With the Abbott scheme (Abbott and Ionescu, 1967) the convective acceleration term is split because the discharge q and the stage h are computed at alternate space steps, as shown in Fig. 4.3.3. Equation (4.3.3) is used to replace the convective acceleration term in (4.4.4) as follows:

$$\frac{\partial}{\partial x}\left(\frac{q^2}{A}\right) = \frac{2q}{A}\frac{\partial q}{\partial x} - \frac{q^2}{A^2}\frac{\partial A}{\partial x} = \frac{2q}{A}\left[br - b\frac{\partial h}{\partial t}\right] - \frac{q^2}{A^2}b\frac{\partial h}{\partial x}. \qquad (4.3.11)$$

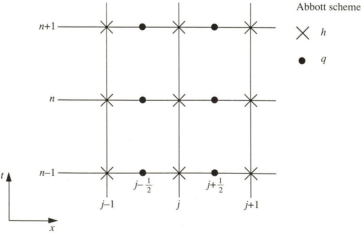

Figure 4.3.3: Abbott scheme mesh.

This results in the momentum equation expressed in the form

$$\frac{\partial q}{\partial t} + \frac{2q}{A}\left[br - b\frac{\partial h}{\partial t}\right] + \left[gA - b\frac{q^2}{A^2}\right]\frac{\partial h}{\partial x} + g\frac{q|q|}{Af^2} = 0 \qquad (4.3.12)$$

using $s_f = q|q|/(A^2f^2)$ where f is a friction parameter, also $\partial h/\partial t = \partial H/\partial t$, $\partial h/\partial x = (\partial H/\partial x) - s$. This illustrates the different forms the St Venant equations can take before discretization.

The Abbott scheme uses the stage h evaluated at space points x_j and the discharge q at space points $x_{j+1/2}$ usually half way between x_j and x_{j+1}. The continuity equation is satisfied as centralized at x_j, $t + \frac{1}{2}\Delta t_n$:

$$\frac{\partial h}{\partial t} \approx \frac{1}{\Delta t_n}(h_j^{n+1} - h_j^n) \qquad (4.3.13a)$$

$$\frac{\partial q}{\partial x} \approx \frac{2}{\Delta x_{j-1} + \Delta x_j}[\theta(q_{j+1/2}^{n+1} - q_{j-1/2}^{n+1}) + (1-\theta)(q_{j+1/2}^n - q_{j-1/2}^n)] \qquad (4.3.13b)$$

with special provision for calculating the width b at junctions to satisfy continuity.

The momentum equation is centralized at $x_{j+1/2}$, $t + \frac{1}{2}\Delta t_n$ using

$$\frac{\partial q}{\partial t} = \frac{1}{\Delta t_n}(q_{j+1/2}^{n+1} - q_{j+1/2}^n)$$

$$\frac{\partial h}{\partial x} = \frac{1}{\Delta x_j}[\theta(h_{j+1}^{n+1} - h_j^{n+1}) + (1-\theta)(h_{j+1}^n - h_j^n)]$$

$$q|q| = |q_{j+1/2}^n|q_{j+1/2}^{n+1}$$

and

$$G = \theta G_{j+1/2}^{n+1} + (1-\theta)G_{j+1/2}^n \qquad (4.3.14)$$

where G is any of the terms q/A, A, q^2b/A^2, and $g/(Af^2)$. The Newton–Raphson iterative method for the non-linear equations (Appendix 2). now gives a system with a tridiagonal matrix which can be solved using the standard algorithm (Conte and de Boor, 1980).

With equal weighting in the difference expressions, $w = \theta = 0.5$ in (4.3.8a) and $\theta = 0.5$ in (4.3.8b), the box and Abbott schemes both have second-order accuracy. It is more difficult to include the boundary conditions and structures where q and h are related with the Abbott scheme but the Abbott scheme has a simpler matrix system to solve. Samuels (1990) claims that the Abbott scheme is less robust near critical flow and hence will need a smaller Δx here than the box scheme.

The next section looks at some limitations on the use of the box scheme.

4.4 Limitations on the use of the box scheme

Samuels and Skeels (1990), taking the St Venant equations together with the general friction law (1.2.4), consider a small perturbation from the steady state and for this show that the box scheme described in section 3.8 is stable for $\theta \geq 0.5$ so long as the Vedernikov number $\mathbf{V}$ is less than or equal to one. The Vedernikov number is given by

$$\mathbf{V} = \frac{m\mathbf{F}A}{nR}\frac{\mathrm{d}R}{\mathrm{d}A} \qquad (4.4.1)$$

where m, n are the parameters in the friction law

$$s_{\mathrm{f}} = \frac{Cu|u|^{n-1}}{R^m}. \qquad (4.4.2)$$

R is the hydraulic radius equal to A/p, where p is the wetted perimeter, u is the mean velocity across the cross-section equal to q/A, and $\mathbf{F}$ is the Froude number of the flow, $\mathbf{F} = u/\sqrt{(gH)}$. When the Vedernikov number $\mathbf{V} > 1$ the nature of the flow changes to rapidly varying unsteady flow where the assumptions made in the formation of the St Venant equations no longer apply and roll waves may form (see French, 1987). Hence this result of Samuels and Skeels links the stability of the box scheme to the physics of the stability of the flow. They give an example of the application of this result to the solution of the flow resulting from a flash flood in an Arabian wadi where the appropriate friction law gives $m = 4/3$, $n = 2$ and the channel is sufficiently wide to make $\mathbf{V} = 2\mathbf{F}/3$ so that $\mathbf{V} \leq 1$ gives $\mathbf{F} \leq 3/2$. The results show that the flow does actually become supercritical for part of the time but without any trouble from numerical oscillations.

The operation of some types of engineering structures can produce a surge in the flow in a channel or pipe. The surge itself corresponds to physical conditions beyond those assumed in the formulation of the St Venant equations. But if there is gradually varied flow either side of the surge then the St Venant box scheme can still be used. Chaudhry and Contractor (1973) recommend using $\theta = 0.6$ although $0.5 \leq \theta < 0.6$ may be used for channels of greater roughness.

Thus with proper verification and caution the box scheme for the St Venant equations may be used for flows which may include conditions beyond those assumed in the original formulation. A practical approach would be to run the program first with $\theta = 0.5$ which gives the most accurate solution, but if this produces oscillations then θ can be gradually increased.

For a comprehensive survey of finite difference methods for one-dimensional surface flow and other topics relating to computational river hydraulics see Cunge *et al.* (1980).

4.5 Methods for two-dimensional surface flow

Many methods for the vertically integrated two dimensions in plan equations for continuity and momentum in surface flow use a combination of ADI (Alternating Direction Implicit) and PC (Predictor–Corrector) methods. PC methods are explained in section 3.2. The following is a simple example to illustrate ADI–PC methods using the parabolic equation

$$\frac{\partial u}{\partial t} = \frac{\partial^2 u}{\partial x^2} + \frac{\partial^2 u}{\partial y^2}. \tag{4.5.1}$$

1(a). Predictor step: eqn (4.5.1) is first approximated over a half time step by the difference scheme to give the predictor values $\hat{u}_{j,k}^{n+1/2}$

$$\frac{\hat{u}_{j,k}^{n+1/2} - u_{j,k}^n}{\Delta t/2} = \frac{\delta^2 x}{\Delta x^2} \hat{u}_{j,k}^{n+1/2} + \frac{\delta^2 y}{\Delta y^2} u_{j,k}^n \tag{4.5.2}$$

where $u_{j,k}^n$ is the approximation to $u(j\Delta x, k\Delta y, n\Delta t)$. Assuming all the values at time level $n\Delta t$ are known, each equation (4.5.2) has just three unknowns, $\hat{u}_{j-1,k}^{n+1/2}, \hat{u}_{j,k}^{n+1/2}, \hat{u}_{j+1,k}^{n+1/2}$, all on the line $y = k\Delta y$, parallel to the x axis and uncoupled from the unknowns on the other such lines. These can be solved separately by the tridiagonal algorithm (Conte and de Boor, 1980) in this case. They could be solved in parallel.

1(b). In the corrector step $u_{j,k}^n$ on the right-hand side of (4.5.2) is replaced by

$$\bar{u}_{j,k}^{n+1/2} = \tfrac{1}{2}(u_{j,k}^n + \hat{u}_{j,k}^{n+1/2}) \tag{4.5.3}$$

and $u_{j,k}^{n+1/2}$ is calculated from

$$\frac{u_{j,k}^{n+1/2} - u_{j,k}^n}{\Delta t/2} = \frac{\delta^2 x}{\Delta x^2} u_{j,k}^{n+1/2} + \frac{\delta^2 y}{\Delta y^2} \bar{u}_{j,k}^{n+1/2}. \tag{4.5.4}$$

2(a). With all the grid values at time $(n + 1/2)\Delta t$ known, the algorithm is completed by solving with a predictor step over the other half time step

$$\frac{\hat{u}_{j,k}^{n+1} - u_{j,k}^{n+1/2}}{\Delta t/2} = \frac{\delta^2 x}{\Delta x^2} u_{j,k}^{n+1/2} + \frac{\delta^2 y}{\Delta y^2} \hat{u}_{j,k}^{n+1} \qquad (4.5.5)$$

along the lines of grid points parallel to the y axis followed by

2(b). the corresponding corrector step.

Thus the values of $u_{j,k}^{n+1}$ at all the grid points are known. The local truncation error is $O(\Delta t^2)$ plus $O(\Delta x^2)$.

The vertically integrated momentum equations can be expressed in the form given by Falconer (1986)

$$\underbrace{\frac{\partial}{\partial t}(q_x)}_{\text{(I)}} + \beta \underbrace{\left[\frac{\partial}{\partial x}(uq_x) + \frac{\partial}{\partial y}(uq_y)\right]}_{\text{(II)}} + \underbrace{gH\frac{\partial h}{\partial x}}_{\text{(III)}} - \underbrace{fvH}_{\text{(IV)}}$$

$$- \varepsilon H \underbrace{\left[2\frac{\partial^2 u}{\partial x^2} + \frac{\partial^2 u}{\partial y^2} + \frac{\partial^2 v}{\partial x \, \partial y}\right]}_{\text{(V)}} + \underbrace{\text{other friction terms}}_{\text{(VI)}} = 0 \quad (4.5.6)$$

and

$$\underbrace{\frac{\partial}{\partial t}(q_y)}_{\text{(I)}} + \beta \underbrace{\left[\frac{\partial}{\partial x}(vq_x) + \frac{\partial}{\partial y}(vq_y)\right]}_{\text{(II)}} + \underbrace{gH\frac{\partial h}{\partial y}}_{\text{(III)}} - \underbrace{fuH}_{\text{(IV)}}$$

$$- \varepsilon H \underbrace{\left[\frac{\partial^2 v}{\partial x^2} + 2\frac{\partial^2 v}{\partial y^2} + \frac{\partial^2 u}{\partial x \, \partial y}\right]}_{\text{(V)}} + \underbrace{\text{other friction terms}}_{\text{(VI)}} = 0 \quad (4.5.7)$$

where β is the correction parameter for non-uniform vertical velocity, f is the Coriolis parameter, ε is the depth mean eddy viscosity, and the friction terms can include bed friction and wind stress. The second-derivative terms from the eddy viscosity make the equations parabolic. (See the ASCE Task Committee (1988) papers on turbulence modelling for a discussion of the equations of surface water flow which can represent the turbulence accurately and are possible to solve numerically.) The continuity equation can be written in the form

$$\frac{\partial h}{\partial t} + \frac{\partial}{\partial x}(q_x) + \frac{\partial}{\partial y}(q_y) = 0. \qquad (4.5.8)$$

These three equations have to be solved for the discharge components, q_x, q_y, and the height of the water surface, h (from which H can be obtained as the height of the base of the flow is assumed known).

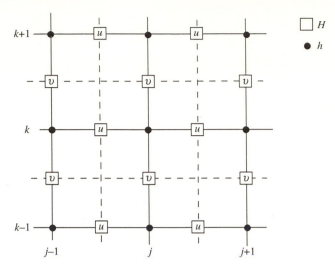

Figure 4.5.1: Falconer scheme—staggered grid.

The following is a summary of the method given in Falconer (1986). This uses a staggered grid with the equations solved for q_x, q_y, h using ADI–PC and provision for upwinding (section 3.7). The grid is shown in Fig. 4.5.1 with q_x, u at points $[(j + 1/2)\Delta x, k\Delta y, (n + 1/2)\Delta t]$, q_y, v at points $[j\Delta x, (k + 1/2)\Delta y, n\Delta t]$, H at points $[(j + 1/2)\Delta x, k\Delta y, n\Delta t]$ and $[j\Delta x, (k + 1/2)\Delta y, n\Delta t]$, and h at points $[j\Delta x, k\Delta y, n\Delta t]$ and $[j\Delta x, k\Delta y, (n + 1/2)\Delta t]$.

The terms in eqn (4.5.6) are numbered so that their treatment can be described in words.

Step 1: for the ADI step in the x direction eqns (4.5.6) and (4.5.8) are used to calculate q_x at points $[(j + 1/2)\Delta x, k\Delta y, (n + 1/2)\Delta t]$ and h at points $[j\Delta x, k\Delta y, (n + 1/2)\Delta t]$, $j = 1, 2, \ldots$. In (4.5.6) the numbered terms are treated as follows:

(I) is represented by

$$\frac{1}{\Delta t}[(q_x)_{j+1/2,k}^{n+1/2} - (q_x)_{j+1/2,k}^{n-1/2}]. \tag{4.5.9}$$

(II)(a) On the Predictor step this term uses q_x values as calculated at time level $n - 1/2$, and if the result of this step is to give values $\hat{q}_x$, (b) these are used in the corrector step to replace the q_x and corresponding u values in term (II) by averages as in (4.5.3).

(III) The pressure gradient term is replaced by a centralized box scheme

$$\frac{1}{2\Delta x} g H_{j+1/2,k}^n [h_{j+1,k}^{n+1/2} - h_{j,k}^{n+1/2} + h_{j+1,k}^{n-1/2} - h_{j,k}^{n-1/2}]. \qquad (4.5.10)$$

(IV) The Coriolis term is represented at time level $n\Delta t$.

(V), (VI) In the bed friction term where the product $u(u^2 + v^2)^{1/2}$ occurs, the first factor is replaced by the average

$$\tfrac{1}{2}(u_{j+1/2,k}^{n+1/2} + u_{j+1/2,k}^{n-1/2})$$

otherwise u is treated by PC as in (II). All other terms are at time level $n\Delta t$.

There is also provision for upstream differencing on the q_y terms. The continuity equation is represented by

$$\frac{1}{\Delta t/2} [h_{j,k}^{n+1/2} - h_{j,k}^n] + \frac{1}{\Delta x} [(q_x)_{j+1/2,k}^{n+1/2} - (q_x)_{j-1/2,k}^{n+1/2}]$$

$$+ \frac{1}{\Delta y} [(q_y)_{j,k+1/2}^n - (q_y)_{j,k-1/2}^n] = 0 \quad (4.5.11)$$

and treated by the PC approach coupled with the momentum equations.

The net result of this combination of staggered grid and averaging is a set of equations for each grid line parallel to the x axis for the unknowns

$$(q_x)_{j-1/2,k}^{n+1/2}, h_{j,k}^{n+1/2}, (q_x)_{j+1/2,k}^{n+1/2}, h_{j+1,k}^{n+1/2}$$

with a four-diagonal matrix.

Step 2: The process is repeated for the other half time step with the unknowns arranged along the grid lines parallel to the y axis. The error is $O(\Delta t^2) + O(\Delta x^2)$ with a square grid.

Weare (1979) discusses the errors arising from the use of ADI in a channel which is not aligned to the grid and says that this problem is inherent in any splitting scheme. His conclusion is that ADI should only be used with Courant numbers less than or equal to two, particularly where flow conditions near an irregular boundary are important.

The theoretical treatment of ADI is restricted to rectangles (see, for instance, Varga, 1962), but this is because, as so often happens, the proofs are easy for this case.

The scheme described by Benque *et al.* (1982) uses a non-staggered grid with a fractional step method. In this scheme the differential operator is split so that the first-derivative advection terms and the second-derivative eddy viscosity diffusion terms are treated separately and then the results combined with the continuity equation as follows:

Step 1: Terms (I) and (II) in both equations (4.5.6) and (4.5.7) are solved for the advection effect using characteristics (Chapter 2) and interpolating back on to the fixed grid. The use of characteristics means that there is no need for upwind differencing.

Step 2: Terms (I) and (V) are used to solve for the diffusion effect using ADI with a double sweep and values from Step 1 where required.

Step 3: This combines the continuity equation (4.5.8) expressed with the θ method (section 3.3) and the remainder of the momentum equations linearized and solved by ADI with iterations. Benque *et al.* say that the number of iterations required for convergence is small for Courant numbers less than or equal to 10. In this two-dimensional problem the Courant number is given by

$$\Delta t \sqrt{(gH)} \left[\frac{1}{\Delta x^2} + \frac{1}{\Delta y^2} \right]^{1/2}.$$

Reeve and Hiley (1992) in their treatment of the depth-integrated equations for tidal flow in shallow water use body- or boundary-fitting coordinates tailored to fit irregular boundaries (Thompson *et al.*, 1982). These are transformed so that a cell in physical space becomes a square in the computation space. The two components of velocity, u, v, and the surface height h are located at the centroids of the elements in computation space and the equations are represented by finite difference expressions used on the transformed equations. The code is described by Burns and Wilkes (1987) who warn against strongly distorted grids in the physical space. This is reminiscent of the situation with the isoparametric elements where it is best not to be too far away from parallelograms (see section 6.7). Reeve and Hiley use a backward difference in time and within each time step solve iteratively as follows:

Step 1: Solve for u, v.

Step 2: Solve for an h correction using results from Step 1.

Step 3: Update, u, v, h, and H.

The wetting and drying in intertidal regions is represented with an algorithm using a wetness parameter to control the mass flow.

Gambolati *et al.* (1990) includes examples of the numerical solution of the equations of three-dimensional surface flow.

5 Finite difference methods applied to groundwater flow

5.1 Introduction

This chapter discusses finite difference methods as used for problems of groundwater flow. Sections 5.2 and 5.3 discuss the elliptic boundary value problems obtained for a confined or unconfined aquifer after vertical integration and averaging. These may include abstraction or recharge wells and stream leakage. Section 5.4 considers the parabolic problems when the aquifers of sections 5.2 and 5.3 have time-dependent flow. Also in section 5.4 are brief descriptions of the IFDM (Integrated Finite Difference Method) for groundwater flow (Narasimhan and Witherspoon, 1976) and SHE (Système Hydrologique Européen) (Abbott et al., 1986) which is a finite difference method for combined surface and subsurface flow. Section 5.5 considers a particular problem of 'noise', i.e. unwanted oscillation that has occurred in the numerical model of an aquifer which includes ephemeral streams.

Chapter 9 contains further examples of finite difference methods applied to water-quality modelling in groundwater flow.

5.2 Steady-state groundwater flow—elliptic boundary value problems

From section 1.7 the steady-state equation for the piezometric head $\phi(x, y)$ in a confined or unconfined aquifer after vertical integration and averaging is

$$-\frac{\partial}{\partial x}\left(T_x \frac{\partial \phi}{\partial x}\right) - \frac{\partial}{\partial y}\left(T_y \frac{\partial \phi}{\partial y}\right) = Q \qquad (5.2.1)$$

where x, y are coordinates in the horizontal plane, Q is a source or sink term (LT^{-1}) which can include wells and/or streams, and T_x, T_y are the transmissivities (dimensions L^2T^{-1}) in the x and y directions resulting from integrating from the base to the top of the saturated layer as shown in section 1.7 and Appendix 1. In the confined aquifer case the top is the top of the confined layer; in the unconfined case the top is taken as the height of the phreatic surface. Values of T_x, T_y can be obtained from site pumping tests (Fig. 1.7.1). Frequently it is considered adequate to take the transmissivity as constant, independent of the piezometric head. A more refined model is made by using eqns (1.7.8b) to include the dependence of the transmissivity on the piezometric head and possibly also the variation in the vertical of the hydraulic conductivity. Connorton and Reed (1978) and Connorton (1980) discuss this with reference to the vertical distribution of fissures in a chalk aquifer.

Equation (5.2.1) holds in the region R of the aquifer. This is an elliptic boundary value problem and boundary conditions are necessary all round the boundary $S = S_1 + S_2$ of R (Fig. 5.2.1). These are usually:

(1) on S_1 'known head' (Dirichlet), alongside a river or lake where the water level is known or where there are borehole measurements; and

(2) on S_2 'known flow' (Neumann); this may be an impermeable boundary or a natural groundwater divide where there is no flow in a direction normal to the boundary. Alternatively the normal flow across the boundary may be known from local conditions. This boundary condition can be implemented by the introduction of a fictitious node as illustrated in section 5.3 but this is not necessary with the integration method as shown in this section.

It is advisable to obtain the finite difference approximation of eqn (5.2.1) by the integration method discussed in section 3.6. This not only ensures a symmetric matrix but also deals with source or sink terms in a simple way. The assumption here is that this is the numerical model of an aquifer which is of such a size that the relative dimensions

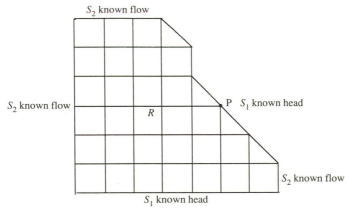

Figure 5.2.1: Aquifer R with boundaries S_1, S_2.

of an abstraction or recharge well are small enough for it to be treated as a point sink or source. The grid is arranged so that any well is at a grid point, (x_w, y_w). The well can then be represented by using the Dirac delta generalized function

$$\delta(x - x_w, y - y_w) \qquad (5.2.2)$$

which has the property that

$$\delta(x - x_w, y - y_w) = 0, \quad x \neq x_w, \quad y \neq y_w \qquad (5.2.3)$$

but

$$\iint\limits_{\Delta R} Q(x, y)\delta(x - x_w, y - y_w)\, dx\, dy = Q(x_w, y_w) \qquad (5.2.4)$$

where the integration is over the region ΔR which includes the point (x_w, y_w) at which the well is situated. The Dirac delta generalized function here has dimension L^{-2} and has the property of filtering out the value of $Q(x, y)$ at the point (x_w, y_w). Thus the wells in the region R can be represented by putting

$$Q = \sum_n \hat{Q}_n \delta(x - x_n, y - y_n) \qquad (5.2.5)$$

where $\hat{Q}_n$ is the rate of abstraction (dimension $L^3 T^{-1}$) from or recharge to the well number n located at (x_n, y_n).

Integrating eqn (5.2.1) round a grid point $(j\Delta x, k\Delta y)$ where there is a well and dividing by $\Delta x \Delta y$, the finite difference result is, from (3.6.11),

$$D_{j,k}\phi_{j,k} - W_{j,k}\phi_{j-1,k} - E_{j,k}\phi_{j+1,k} - N_{j,k}\phi_{j,k+1}$$
$$- S_{j,k}\phi_{j,k-1} = \hat{Q}_n(j\Delta x, k\Delta y)/(\Delta x \Delta y) \qquad (5.2.6)$$

where

$$W_{j,k} = \frac{1}{\Delta x^2} T_{x_{j-1/2,k}}, \qquad E_{j,k} = \frac{1}{\Delta x^2} T_{x_{j+1/2,k}}$$

$$N_{j,k} = \frac{1}{\Delta y^2} T_{y_{j,k+1/2}}, \qquad S_{j,k} = \frac{1}{\Delta y^2} T_{y_{j,k-1/2}}$$

$$D_{j,k} = W_{j,k} + E_{j,k} + N_{j,k} + S_{j,k}. \qquad (5.2.7)$$

Rushton (1981) proposes a correction factor to give an improved value of the piezometric head at a well number n:

$$\phi'_n = \phi_n + a\hat{Q}_n \qquad (5.2.8)$$

where ϕ_n is the piezometric head calculated by the finite difference equation and a is a parameter to be deduced from site observation.

A convenient way to represent stream leakage is with a series of Dirac delta generalized functions at grid points along the stream bed. Stream leakage with special reference to problems with the numerical solution with ephemeral streams is discussed in section 5.5.

Of course if this is a steady-state problem there should be a mass balance:

$$\text{flow in} = \text{flow out}.$$

This is calculated from the values of the discharge

$$q_P = -\left(T_x \frac{\partial \phi}{\partial x} n_x + T_y \frac{\partial \phi}{\partial y} n_y \right) \delta S \qquad (5.2.9)$$

at nodes P on the fixed head boundary S_1 (Fig. 5.2.1), where (n_x, n_y) is the unit vector representing the outward normal to the boundary S_1 at P and δS is the length of S_1 in the neighbourhood of P (i.e. half way to the next node each way on S_1). The boundary is taken to be on a grid line or to consist of lines joining grid points as shown in Fig. 5.2.1. The values of q_P added together over all points P on S and combined with the difference between abstraction and recharge rates represented by Q should equal zero but in practice there will be an error. The size of this error can be taken as an indicator to show whether the mesh size $(\Delta x, \Delta y)$ is small enough.

Note that this operation is equivalent to integrating eqn (5.2.1) over the whole region R:

$$-\iint_R \left\{ \frac{\partial}{\partial x}\left(T_x \frac{\partial \phi}{\partial x} \right) + \frac{\partial}{\partial y}\left(T_y \frac{\partial \phi}{\partial y} \right) \right\} dx\, dy = \iint_R Q\, dx\, dy \qquad (5.2.10)$$

and using the Gauss divergence theorem (Spencer *et al.*, 1977) to give the mass-balance equation

$$-\int_{S_1+S_2} \left(T_x \frac{\partial \phi}{\partial x} n_x + T_y \frac{\partial \phi}{\partial y} n_y \right) dS = [Q] \qquad (5.2.11)$$

(the net inflow from the wells), i.e. if there is no flow across S_2,

$$-\int_{S_1} \left(T_x \frac{\partial \phi}{\partial x} n_x + T_y \frac{\partial \phi}{\partial y} n_y \right) dS = [Q]. \qquad (5.2.12)$$

It is the left-hand side of (5.2.12) which is approximated by summing the contributions from (5.2.9).

5.3 Boundary conditions

As stated in section 5.2 the boundary is taken to be on a grid line or to consist of lines joining grid points as shown in Fig. 5.2.1. At known-head grid points $(j\Delta x, k\Delta y)$ on S_1 the known value, ϕ_{jk}, is written into the formula (5.2.6). At known-flow grid points P at $(j\Delta x, k\Delta y)$ on S_2 the value of ϕ is unknown. If this point is on part of the boundary parallel to an axis as in Fig. 5.3.1 the integration region ΔR joining the points APBCD is now half the previous size as shown. The contribution from the line integral along APB is given by the known flow. Hence the integration method produces the equation for $\phi_{j,k}$ as

$$(\tfrac{1}{2}N_{j,k} + \tfrac{1}{2}S_{j,k} + W_{j,k})\phi_{j,k} - W_{j,k}\phi_{j-1,k}$$

$$- \tfrac{1}{2}N_{j,k}\phi_{j,k+1} - \tfrac{1}{2}S_{j,k}\phi_{j,k-1} = \text{known flow across AB}/(\Delta x\Delta y) \quad (5.3.1)$$

where $N_{j,k}$, $S_{j,k}$, $W_{j,k}$ are as given in (5.2.7).

If part of the S_2 boundary is joining $(j\Delta x, k\Delta y)$ to $((j-1)\Delta x, (k+1)\Delta y)$ and $((j+1)\Delta x, (k-1)\Delta y)$ as shown in Fig. 5.3.2, the integration area ΔR is now the triangle ABC. Again known flow perpendicular to the boundary means a known contribution to the line integral along AB and the resulting equation for $\phi_{j,k}$ is

$$(S_{j,k} + W_{j,k})\phi_{j,k} - S_{j,k}\phi_{j,k-1} - W_{j,k}\phi_{j-1,k} = \text{known flow across AB}/(\Delta x\Delta y).$$

$$(5.3.2)$$

(Note: it is important to check dimensions and the sign of the

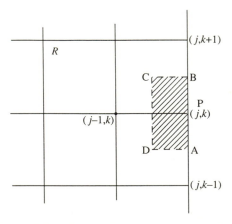

Figure 5.3.1: Known-flow point on boundary.

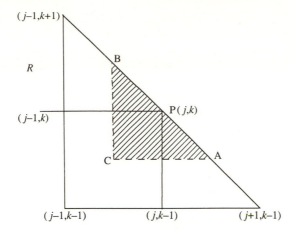

Figure 5.3.2: Known-flow point on boundary, triangular ΔR.

right-hand-side terms in (5.3.1) and (5.3.2); water piling up just inside the boundary shows the sign is wrong.) Curved boundaries or boundary conditions with given non-zero flow are more easily dealt with by finite elements and so will not be considered further here. Also only a regular mesh has been considered here so this difference scheme is centred and the local truncation error is $O(\Delta x^2) + O(\Delta y^2)$ as shown in (3.6.8). If an irregular mesh is used the difference scheme is no longer centred and the local truncation error has only first-order terms.

In the solution of the algebraic equations from the difference scheme produced by the integration method just described advantage can be taken of the symmetry of the matrix. Methods of solution of the algebraic equation are listed in Appendix 2.

5.4 Time-dependent groundwater flow—parabolic problems

As shown in section 1.7 for time-dependent vertically integrated and averaged groundwater flow eqn (5.2.1) is replaced by the parabolic equation

$$\frac{\partial}{\partial x}\left(T_x \frac{\partial \phi}{\partial x}\right) + \frac{\partial}{\partial y}\left(T_y \frac{\partial \phi}{\partial y}\right) = S \frac{\partial \phi}{\partial t} - Q \qquad (5.4.1)$$

$(t > 0)$, where S is the effective storativity.

If the aquifer is represented by the region R in Fig. 5.2.1 the boundary conditions of the two types, (1) known head on S_1 and (2) known flow on S_2, must now be given for all time $t > 0$. It is also necessary to know the initial values of ϕ all over the region because this is now an initial value problem (Table 1.1.1).

The initial condition $\phi(x, y, 0)$ is usually obtained from (a) a steady-state solution with ϕ constant everywhere so that there is no flow, or (b) a 'dynamic balance' steady state with flow in equal to flow out.

The finite difference approximation to $\phi(j\Delta x, k\Delta y, n\Delta t)$ is denoted by $\phi_{j,k}^n$. Equation (5.4.1) is discretized in space in the same way as (5.2.1). Suppose the latter result is written as

$$L_{\Delta x, \Delta y} T\phi_{j,k} + \hat{Q}_{j,k} = 0 \qquad (5.4.2)$$

(taking $T_x = T_y = T$, constant, for simplicity). Then as shown for the one-space-dimensional parabolic problem in section 3.7, the corresponding θ method applied to (5.4.1) is (see eqn (3.7.7))

$$\frac{S(\phi_{j,k}^{n+1} - \phi_{j,k}^n)}{T\Delta t} = \theta L_{\Delta x, \Delta y} \phi_{j,k}^{n+1} + (1 - \theta) L_{\Delta x, \Delta y} \phi_{j,k}^n + \hat{Q}_{j,k}. \qquad (5.4.3)$$

For stability analysis the source or sink term $\hat{Q}$ can be left out. The stability of the numerical method (5.4.3) is then analysed by the von Neumann method (section 3.7) looking at the response to a Fourier mode:

$$\phi_{j,k}^n = \zeta^n \alpha_{p,q} \exp(ijp\Delta x) \exp(ikq\Delta y)$$

where $i = \sqrt{-1}$ and p, q are integers. This gives

$$L_{\Delta x, \Delta y} \phi_{j,k}^n = -4\zeta^n \left(\frac{s_1^2}{\Delta x^2} + \frac{s_2^2}{\Delta y^2} \right) \alpha_{p,q} \exp(ijp\Delta x) \exp(ikq\Delta y) \qquad (5.4.4)$$

where $s_1 = \sin(p\Delta x/2)$, $s_2 = \sin(q\Delta y/2)$.

Put $A_1 = s_1^2/\Delta x^2$, $A_2 = s_2^2/\Delta y^2$, substitute into (5.4.3) with $\hat{Q} = 0$, and cancel $\zeta^n \alpha_{p,q} \exp(ijp\Delta x) \exp(ikq\Delta y)$ to give the equation

$$S(\zeta - 1) = -4\Delta t T(A_1 + A_2)(\theta \zeta + 1 - \theta). \qquad (5.4.5)$$

Rearranging (5.4.5) gives the equation for the amplification factor ζ as

$$\zeta = \frac{S - 4\Delta t T(1 - \theta)(A_1 + A_2)}{S + 4\Delta t T\theta(A_1 + A_2)}. \qquad (5.4.6)$$

As before for a decaying numerical solution it is required to have $|\zeta| < 1$, i.e. $-1 < \zeta < 1$, and this condition is satisfied if

$$S > 2\Delta t T(1 - 2\theta)(A_1 + A_2). \qquad (5.4.7)$$

(Note: S is dimensionless, T has dimensions L^2T^{-1}, and A_1, A_2 are L^{-2}; hence the right-hand side is also dimensionless.) Condition (5.4.7) is satisfied for any value of Δt if $\theta \geq 0.5$, i.e. there is then unconditional stability of the scheme. If $\theta < 0.5$ the scheme is stable if

$$\Delta t < \frac{S}{2T(1 - 2\theta)(A_1 + A_2)}.$$

Since the maximum value that a sine can have is unity this means that for all possible Fourier modes the scheme is stable if

$$\Delta t < \frac{S}{2T(1 - 2\theta)} \left(\frac{1}{\Delta x^2} + \frac{1}{\Delta y^2} \right)^{-1}. \tag{5.4.8}$$

In particular for the explicit method given by $\theta = 0$ and with equal mesh sizes $\Delta x = \Delta y$, the condition on the time step for stability is

$$\frac{T\Delta t}{S\Delta x^2} < \tfrac{1}{4}. \tag{5.4.9}$$

Note that the principal part of the truncation error of the scheme (5.4.3) as shown in section 3.2 contains terms in Δt^2 (for $\theta = 0.5$), or Δt (for $\theta \neq 0.5$), times a derivative of the solution which may be large when the solution is changing rapidly. An adjustable time step is required to control the error, e.g. if the maximum change in the solution at any point is greater than some tolerance the time step is repeated with Δt halved.

The integrated finite difference method

MacNeal (1953) uses the concept of mesh subregions. A number of nodes are chosen in the region and on its boundary. Their position is arbitrary except that when joined up into triangles these have angles less than or equal to 90°. This is so that the circumcentre of each triangle, which is the point of intersection of the perpendicular bisectors of the sides, does not lie outside the triangle. Each node such as A (x_j, y_k) in Fig. 5.4.1 is associated with a subregion R_{jk} which is the polygon 12345 connecting the circumcentres of the triangles of which A is a vertex. The differential equation (5.4.1) can be integrated over the subregion R_{jk} and the Gauss divergence theorem used on the second-derivative terms so that these are replaced by a sum of fluxes across the sides of the polygon:

$$\sum_j T_{Aj}(\phi_A - \phi_{Bj}) l_{ABj}/(AB_j)$$

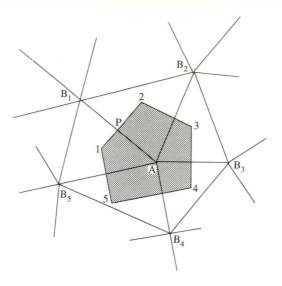

Figure 5.4.1: Integrated finite difference method—example.

where l_{AB_j} is the length of the side of the polygon perpendicular to AB_j and T_{A_j} is the harmonic mean of the transmissivities in the subregions based on nodes A and B_j. (This can be demonstrated as follows. Continuity of flow at a point such as P in Fig. 5.4.1 requires that

$$T_A(\phi_A - \phi_P)/PA = T_{B_1}(\phi_P - \phi_{B_1})/B_1P.$$

With $PA = B_1P$ this means that

$$\phi_P = (T_A\phi_A + T_{B_1}\phi_{B_1})/(T_A + T_{B_1}).$$

Thus if the appropriate mean value of the transmissivity is $\bar{T}$ so as to give

$$\bar{T}(\phi_A - \phi_{B_1})/AB_1 = T_A(\phi_A - \phi_P)/PA$$

with $AB_1 = 2PA$, substituting for ϕ_P gives

$$\bar{T}(\phi_A - \phi_{B_1}) = 2T_A T_{B_1}(\phi_A - \phi_{B_1})/(T_A + T_{B_1})$$

thus

$$\bar{T}^{-1} = \tfrac{1}{2}(T_A^{-1} + T_{B_1}^{-1})$$

i.e. $\bar{T}$ is the harmonic mean of T_A and T_{B_1}.)

The result is a system of simultaneous equations of the same form as that obtained by the integration method in section 3.6. The matrix is symmetric and positive definite (see Varga (1962) for further details). Smaller elements can be taken in parts of the region where higher accuracy is required. This method is thus an extension of the integration method described in section 3.6 and also it resembles the finite element method described in Chapter 6. It is the basis of the integrated finite difference method for groundwater flow introduced by Narasimhan and Witherspoon (1976); in this form the lines joining the nodes must be perpendicular to but not necessarily the bisectors of the sides of the subregion. This method can also be extended to three-dimensional groundwater flow problems.

SHE (Abbott *et al.*, 1986) is a finite difference model of surface and subsurface flow with allowance for movement between the surface and the unsaturated and saturated layers. The catchment area is covered by a square grid with a vertical column of horizontal layers beneath each grid square. The overland flow is represented by a simplification of the St Venant equations into the form

$$\frac{\partial h}{\partial t} + \frac{\partial}{\partial x}(uh) + \frac{\partial}{\partial y}(vh) = r \tag{5.4.10}$$

where $\partial h/\partial x = s_x - s_{fx}$, $\partial h/\partial y = s_y - s_{fy}$, and s_x, s_y, s_{fx}, s_{fy} are the ground and friction slopes in the x and y directions respectively; $u = u(h)$ and $v = v(h)$ are obtained from the Manning friction law (section 1.2). Equation (5.4.10) is solved by an explicit time-stepping scheme.

The flow in the unsaturated zone between the ground surface and the phreatic surface is taken to be vertical flow only given by an equation of the form

$$\frac{\partial \theta}{\partial \psi} \frac{\partial \psi}{\partial t} = \frac{\partial}{\partial z}\left(K_z \frac{\partial \psi}{\partial z}\right) + \frac{\partial}{\partial z}(K_z) + Q \tag{5.4.11}$$

where $\psi(\theta)$ and $K_z(z, \theta)$ are obtained from known functional relationships. This is solved by an implicit finite difference scheme with iteration to allow for the non-linearity.

The vertically integrated and averaged equation for the flow in the saturated layer, similar to (5.4.1), is solved by an ADI non-iterative method.

Storm (1991) also discusses the numerical modelling of the interaction between surface and subsurface flow.

The SHE model is extended to deal with contaminant and sediment transport by Bathurst and Purnama (1991).

5.5 Noise problems with ephemeral streams

A particular problem in the numerical modelling of chalk and limestone
aquifers is that of representing the effect of ephemeral streams. These
are streams which have little or no storage capacity; they flow and
take water away from the aquifer when the piezometric head is above
a certain level (usually the stream bed level) and when the piezometric
head is below this level the stream dries up. The chalk and limestone
aquifers here are assumed to be governed by Darcy flow properties;
see Bear (1979) for a reference to non-Darcy flow.

Putting $\phi = h$, the piezometric head, in eqn. (5.4.1), the relevant
equation here is

$$\frac{\partial}{\partial x}\left(T_x\frac{\partial h}{\partial x}\right) + \frac{\partial}{\partial y}\left(T_y\frac{\partial h}{\partial y}\right) = S\frac{\partial h}{\partial t} + Q \qquad (5.5.1)$$

where

$$Q = \sum_N Q_s\delta(x - x_s, y - y_s) \qquad (5.5.2)$$

is a sum of sink terms at the stream nodes using the Dirac delta
generalized function as before. Connorton and Wood (1983) take the
ephemeral stream leakage as represented by

$$Q_s = \begin{cases} K_{sl}(x, y)(h - h_s) & h \geq h_s \\ 0 & h < h_s \end{cases} \qquad (5.5.3)$$

(see Fig. 5.5.1) where $K_{sl}(x, y)$ is the stream leakage parameter.

In a previous paper Connorton and Hanson (1978) when attempting

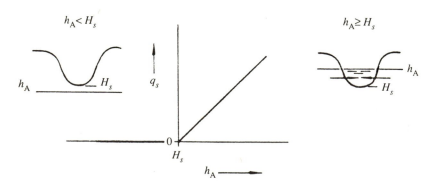

Figure 5.5.1: Graphical representation of stream leakage.

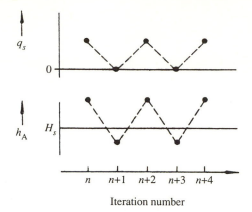

Figure 5.5.2: Sustained oscillation at a stream node.

to solve this problem using finite differences and the SOR (Successive Over Relaxation) iteration method (Appendix 2) had trouble with noise in the form of sustained oscillations such that the iteration never converged as shown in Fig. 5.5.2. The cure for this can be demonstrated on the steady-state form of eqn (5.5.1).

Put $h_s H = h$, where h_s is supposed constant here for simplicity, and $T_x = T_y = T(x, y)$ in the steady-state form of the equation; the result is

$$\frac{\partial}{\partial x}\left(T\frac{\partial H}{\partial x}\right) + \frac{\partial}{\partial y}\left(T\frac{\partial H}{\partial y}\right) = K_{sl}(H - 1). \qquad (5.5.4)$$

Take a square mesh, $\Delta x = \Delta y$, and use the integration method (section 3.6) to produce the equation

$$\frac{1}{4T}(-T_{j,k-1/2}H_{j,k-1} - T_{j-1/2,k}H_{j-1,k} + 4TH_{j,k}$$

$$- T_{j+1/2,k}H_{j+1,k} - T_{j,k+1/2}H_{j,k+1}) + \left[K_{sl}\frac{\Delta x b}{4T}H_{j,k}\right] = \left[K_{sl}\frac{\Delta x b}{4T}\right] \qquad (5.5.5)$$

where b is the local average width of the stream and

$$4T = T_{j,k-1/2} + T_{j-1/2,k} + T_{j,k+1/2} + T_{j+1/2,k} \qquad (5.5.6)$$

so that T is the local average transmissivity. Suppose $b < \Delta x$ and that the stream runs along a mesh line for further simplicity. The terms in square brackets in (5.5.5) are only there in the equation when $(j\Delta x, k\Delta y)$ is a stream node.

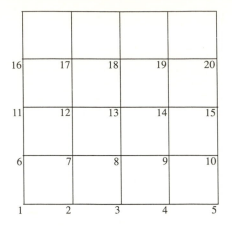

Figure 5.5.3: Node numbering.

Put

$$B = K_{sl} \frac{\Delta x b}{4T}. \tag{5.5.7}$$

This is the key dimensionless parameter here. The larger the value of B the greater the likelihood of noise problems. In a large aquifer model it is not possible to reduce B by reducing Δx. In the Lambourn aquifer, for example, B can have values up to 10^4.

Consider the nodes numbered so that the values of H form a one-dimensional array $\{H_l\}$ as illustrated in Fig. 5.5.3 and suppose the matrix of the system of equations representing the non-bracketed terms in (5.5.5) has entries $a_{l,m}$. Then the SOR algorithm (Appendix 2) applied to the solution of the system of equations (5.5.5) makes the lth equation appear at the kth iteration as

$$H_l^{k+1} = (1 - \omega)H_l^k + \frac{\omega}{1 + B}\left[f_l + B - \sum_{m<l} a_{l,m}H_m^{k+1} - \sum_{m>l} a_{l,m}H_m^k \right] \tag{5.5.8}$$

where $H_l^k \geq 1$ and l is the number of a stream node, or

$$H_l^{k+1} = (1 - \omega)H_l^k + \omega\left[f_l - \sum_{m<l} a_{l,m}H_m^{k+1} - \sum_{m>l} a_{l,m}H_m^k \right] \tag{5.5.9}$$

otherwise. Here ω is the SOR parameter and f_l represents the possible contribution from a known-head node.

The iteration is started with $H_l^0 < 1$ for all l, i.e. the piezometric head is supposed below the stream bed level so that eqn (5.5.9) is used. The values of H_l^k increase until there is some stream node s where

$H_s^k \geq 1$. If $\hat{H}_s^{k+1}$ is the value which would be given if the next step did not include the extra terms, this is given by the equation

$$\hat{H}_s^{k+1} = (1 - \omega)H_s^k + \omega\left[f_s - \sum_{m<s} a_{s,m}H_m^{k+1} - \sum_{m>s} a_{s,m}H_m^k \right]. \quad (5.5.10)$$

But what is actually computed is

$$H_s^{k+1} = \frac{B(1 - \omega)}{1 + B} H_s^k + \frac{1}{1 + B} \hat{H}_s^{k+1} + \frac{\omega B}{1 + B}. \quad (5.5.11)$$

Now it is known that $H_s^k = 1 + \varepsilon_s^k$, where $\varepsilon_s^k \geq 0$. Suppose $\hat{H}_s^{k+1} = 1 + \hat{\varepsilon}_s^{k+1}$. Then substituting into eqn (5.5.11) gives

$$H_s^{k+1} = 1 - \frac{(\omega - 1)}{1 + B} B\varepsilon_s^k + \frac{1}{1 + B} \hat{\varepsilon}_s^{k+1}. \quad (5.5.12)$$

The right-hand side of eqn (5.5.12) shows that

$$H_s^{k+1} < 1 \quad \text{if} \quad (\omega - 1)B\varepsilon_s^k > \hat{\varepsilon}_s^{k+1}. \quad (5.5.13)$$

The SOR parameter ω is taken as 1.6 by Connorton and Hanson (1978) and the numerical value of B can be up to 10^4 so condition (5.5.13) is very likely to be true. Thus a persistent oscillation can arise with $H_s^{k-1} < 1$, $H_s^k > 1$, $H_s^{k+1} < 1$, etc., with the extra terms brought in at alternate iterations. The cure is to insert an extra step in the algorithm so as to take an average

$$\hat{H}_l^k = \tfrac{1}{2}(H_l^k + \hat{H}_l^{k-1}) \quad (5.5.14)$$

followed by

$$H_l^{k+1} = (1 - \omega)\hat{H}_l + \frac{\omega}{1 + B}\left[f_l - \sum_{l<m} a_{l,m}H_m^{k+1} - \sum_{l>m} a_{l,m}\hat{H}_m^k + B \right] \quad (5.5.15)$$

if $\hat{H}_l^k \geq 1$ and l is a stream node, or

$$H_l^{k+1} = (1 - \omega)\hat{H}_l^k + \omega\left[f_l - \sum_{l<m} a_{l,m}H_m^{k+1} - \sum_{l>m} a_{l,m}\hat{H}_m^k \right] \quad (5.5.16)$$

otherwise.

If $\mathbf{L}$ is the SOR iteration matrix corresponding to the set of equations (5.5.8), (5.5.9), $\mathbf{H}^k$, $\hat{\mathbf{H}}^k$ are the vectors of nodal values H_l^k, $\hat{H}_l^k$ at the kth iteration, and $\bar{\mathbf{H}}$ is the vector of nodal values to which the iteration converges, then it follows that

$$\mathbf{H}^{k+1} - \bar{\mathbf{H}} = \mathbf{L}(\hat{\mathbf{H}}^k - \bar{\mathbf{H}}). \quad (5.5.17)$$

Substituting for $\mathbf{H}^{k+1} = 2\hat{\mathbf{H}}^{k+1} - \hat{\mathbf{H}}^k$ from the averaging step (5.5.14) then gives

$$\hat{\mathbf{H}}^{k+1} - \bar{H} = \tfrac{1}{2}(\mathbf{I} + \mathbf{L})(\hat{\mathbf{H}}^k - \bar{\mathbf{H}}). \qquad (5.5.18)$$

Thus the iteration is still convergent because if λ is an eigenvalue of $\mathbf{L}$, the SOR parameter has been chosen to make $\lambda < 1$ and in the new iteration matrix the corresponding eigenvalues are $\tfrac{1}{2}(1 + \lambda)$ which must also be less than unity.

The iteration is continued with the values of $\hat{H}_s$ increasing until $\hat{H}_s^k = 1 + \hat{\varepsilon}_s^k$, $\hat{\varepsilon}_s^k \geq 0$. Then the effect of the averaging step (5.5.14) is to give

$$\hat{H}_s^{k+1} = 1 + \frac{1}{2}\frac{[1 + B(2 - \omega)]}{1 + B}\hat{\varepsilon}_s^k + \frac{1}{2(1 + B)}\hat{\varepsilon}_s^{k+1}. \qquad (5.5.19)$$

Even if $\hat{\varepsilon}_s^{k+1}$ is negative the larger factor multiplying the $\hat{\varepsilon}_s^k$ term is positive. There may be some small oscillations at the start but the results settle down so that $\hat{H}_s^k > 1$ is followed by $\hat{H}_s^{k+1} > 1$ also and the iteration converges.

The formula for $\bar{\omega}$, the optimum value of the SOR parameter (Appendix 2), is

$$\bar{\omega} = \frac{2}{1 + \sqrt{(1 - \mu^2)}}$$

where μ is the spectral radius of the corresponding Jacobi matrix (Conte and de Boor, 1980). The spectral radius of the SOR matrix is then $\lambda_1 = \bar{\omega} - 1$. The effect of the averaging step is to make the spectral radius

$$\lambda_{\mathrm{av}} = \tfrac{1}{2}(1 + \lambda_1) = \tfrac{1}{2}\bar{\omega}. \qquad (5.5.20)$$

Figure 5.5.4 shows the graph of the spectral radius of the

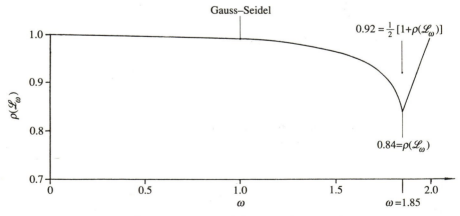

Figure 5.5.4: Spectral radius of the iteration matrix versus ω.

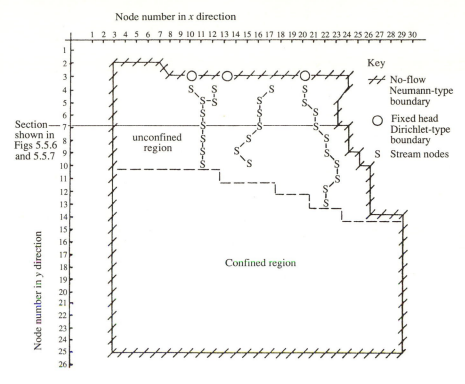

Figure 5.5.5: Finite difference grid for Cotswolds model.

SOR iteration matrix against ω for the numerical model of the Cotswold limestone aquifer in Fig. 5.5.5. The effect of the averaging here is to make the spectral radius of the iteration matrix equal to 0.92.

The averaging procedure described above is applied by Connorton and Wood (1983) to the Cotswold aquifer shown in Fig. 5.5.5. The model includes a confined and an unconfined region with three ephemeral streams. With the transmissivity set to $500\,\mathrm{m}^2$ per day satisfactory results are given when using ordinary SOR. But when the transmissivity is reduced to $100\,\mathrm{m}^2$ per day thereby increasing the value of the parameter B, ordinary SOR gives sustained oscillations as shown in Fig. 5.5.6. These oscillations occur globally but are worst in the neighbourhood of streams. They are still there after 900 iterations. With the averaging method and optimized $\bar{\omega} = 1.85$ there is convergence after 100 iterations (Fig. 5.5.7).

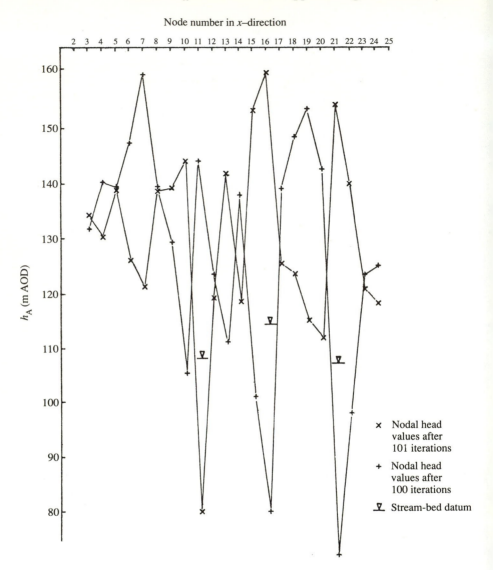

Figure 5.5.6: Section along row 7 (Fig. 5.5.5) at successive iterations (100, 101) without averaging.

This example illustrates the use of averaging to smoothe out persistent spurious oscillations. However, it is important to be aware of the cause of oscillations as averaging is not always the answer.

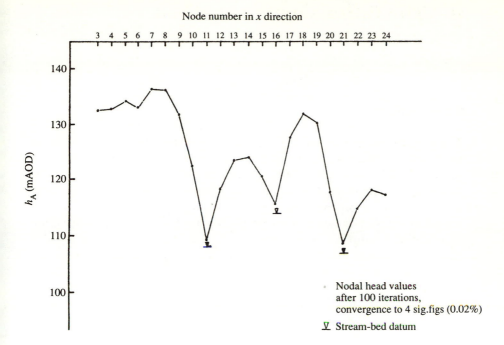

Figure 5.5.7: Section along row 7 after 100 iterations (Fig. 5.5.5) with averaging.

6 The finite element method

6.1 Introduction

The finite element method (FEM) is not recommended in this book for the solution of surface flow problems because finite difference methods are easier for these. The FEM is introduced here as applied to the solution of subsurface flow, i.e. for the spatial discretization of elliptic and parabolic problems (Chapter 1). At first it is assumed that the region concerned is two dimensional. The basic idea is that this region in which the solution is required is divided into triangles or quadrilaterals and in each of these a simple polynomial approximation to the exact solution is taken. The simplest is that of a piecewise linear approximation on triangles. The result of this can be visualized as a polyhedral surface of triangular flat plates approximating a smooth curved surface. The edges of the triangular flat plates projected on to an intersecting plane give the elements into which the region in the plane has been divided (what happens if the region has curved boundaries is considered later—section 6.4). With this image it is clear that the error in the piecewise linear approximation will be proportional to the curvature (i.e. the second derivative) of the surface and that where the curvature is greatest there should be smaller elements.

In this section the FEM is introduced in this form of piecewise linear approximations on triangle elements for the solution of linear problems of groundwater flow. This is used to explain the general procedure whereby the differential equation plus boundary conditions is replaced by a system of simultaneous linear equations. Section 6.2 gives the details of the kinds of matrices which appear in these equations. Section 6.3 describes other kinds of element which can be useful in subsurface flow problems: six-node triangles, four- and eight-node squares, and three-dimensional elements. Section 6.4 introduces the very powerful idea of isoparametric elements which enable the program to use general quadrilaterals and to cope with curved boundaries. Section 6.5 discusses the accuracy of the FEM solution. In a finite element package there is usually a facility for choosing the numerical integration to be used and this is covered in section 6.6. Section 6.7 gives an invaluable method for testing the program called the patch test.

It is convenient to start with the steady-state problem of vertically

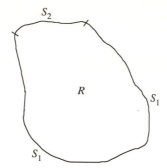

Figure 6.1.1: Region R with boundaries S_1, S_2.

integrated flow as in eqn (1.7.5):

$$-T\frac{\partial^2 \phi}{\partial x^2} - T\frac{\partial^2 \phi}{\partial y^2} = Q \qquad (6.1.1)$$

with constant transmissivity T in an aquifer region R (Fig. 6.1.1) with boundary conditions

$$\phi = g(x, y) \text{ on } S_1 \qquad (6.1.2)$$

the 'known-head' Dirichlet boundary condition, and

$$T\frac{\partial \phi}{\partial n} = \beta(x, y) \text{ on } S_2 \qquad (6.1.3)$$

the 'known-flow' Neumann boundary condition. If this is

$$T\frac{\partial \phi}{\partial n} = 0 \qquad (6.1.4)$$

this is the 'no-flow' condition. Here $\partial\phi/\partial n$ is the gradient of ϕ in the direction of the outward normal to the region R at a point on S_2, i.e. if the outward normal is represented by the unit vector (n_x, n_y)

$$\frac{\partial \phi}{\partial n} = \frac{\partial \phi}{\partial x} n_x + \frac{\partial \phi}{\partial y} n_y.$$

The FEM is a particular development of the Rayleigh–Ritz method where an approximate solution of the problem is constructed in the form

$$\phi \approx \sum_{j=1}^{J} \phi_j N_j(x, y) \qquad (6.1.5)$$

where the $N_j(x, y)$ are basis functions and the ϕ_j are parameters. In the (a)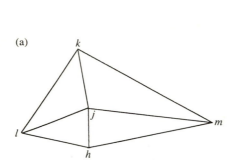

(b)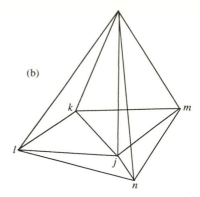

Figure 6.1.2: (a) Plan view of node j and neighbourhood; (b) $N_j(x, y)$ pyramid on base *lkmn*.

original Rayleigh–Ritz method the basis functions $N_j(x, y)$ are global Fourier series terms. In the FEM the region R is divided into elements which in the simplest case are triangles (Fig. 6.1.2). The vertices of these triangles are called nodes and the $N_j(x, y)$ are now local basis functions such that the approximation is piecewise linear and the parameters ϕ_j are the approximate values of ϕ at the nodes. This is achieved by making the basis functions satisfy the conditions:

$$N_j(x, y) = \begin{cases} 1 & \text{at nodes } j, \text{ i.e. at } (x_j, y_j) \\ 0 & \text{at all other nodes} \end{cases} \tag{6.1.6}$$

and $N_j(x, y) \equiv 0$ outside the neighbourhood of node j.

Focus on a node j, (x_j, y_j), to get a picture of the basis function $N(x, y)$ (Fig. 6.1.2). Suppose node j is the common vertex of four triangular elements whose other vertices are the nodes l, k, m, n in the plan view shown in Fig. 6.1.2(a)). The basis function $N_j(x, y)$ satisfies

$$N_j(x_j, y_j) = 1$$

and decreases linearly to zero at the neighbouring nodes l, k, m, n and elsewhere is identically zero. The result is a pyramid of height unity and base outlined by the nodes l, k, m, n which is the neighbourhood of node j shown in Fig. 6.1.2(b).

Focus on an element, triangle jkl, to get a picture of the local approximation to $\phi(x, y)$. Within the triangle jkl the function $\phi(x, y)$ is approximated by

$$\phi^h(x, y) = \phi_j N_j(x, y) + \phi_k N_k(x, y) + \phi_l N_l(x, y) \tag{6.1.7}$$

where the superscript h refers to the size of the elements in a general way. The 'size' of an element, h, e.g. the longest side, is essentially a fraction: the ratio of the actual size of the element to the actual size of the region R. In the right-hand side of (6.1.7) each $N_i(x, y)$, $i = j$, k, l, is linear so that the combined effect is a function $\phi^h(x, y)$ which represents a plane with heights ϕ_j, ϕ_k, ϕ_l at the nodes j, k, l respectively and no other basis function $N_m(x, y)$, $m \neq j, k, l$, can affect this triangle. The net effect is a piecewise linear approximation over the whole region R.

Note that

$$\sum_{j=1}^{J} N_j(x, y) = 1 \qquad (6.1.8)$$

everywhere in the region R because if $\phi_j = 1$ for all j the result must be a plane of height unity. This result is sometimes useful in programming.

It is no use substituting the piecewise linear approximation (6.1.5) constructed as above directly into the left-hand side of eqn (6.1.1) because the second derivative of a linear function is zero. It is necessary first to obtain a more suitable form of eqn (6.1.1). To do this eqn (6.1.1) is multiplied by a 'test function' or 'weight' $w(x, y)$ and integrated over the region R:

$$-T \iint_R \left[\frac{\partial^2 \phi}{\partial x^2} + \frac{\partial^2 \phi}{\partial y^2} \right] w(x, y) \, dx \, dy = \iint_R Q w(x, y) \, dx \, dy. \qquad (6.1.9)$$

The integrand on the left of (6.1.9) is $w(x, y) \nabla^2 \phi$. Integrate this by parts using the product differentiation rule, i.e. (omitting arguments for simplicity)

$$\nabla(w \nabla \phi) = \nabla w \cdot \nabla \phi + w \nabla^2 \phi. \qquad (6.1.10)$$

Substituting into (6.1.9) gives

$$-\iint_R [\nabla(w \nabla \phi) - \nabla w \cdot \nabla \phi] \, dx \, dy = \iint_R Q w \, dx \, dy. \qquad (6.1.11)$$

Then the Gauss divergence theorem (Spencer *et al.*, 1977) can be used on the first term on the left of (6.1.11) to give

$$\iint_R \nabla(w \nabla \phi) \, dx \, dy = \int_{S_1 + S_2} w \frac{\partial \phi}{\partial n} \, dS \qquad (6.1.12)$$

or

$$\iint_R \mathbf{V}(w\nabla\phi)\,\mathrm{d}x\,\mathrm{d}y = \int_{S_1+S_2} w\nabla\phi\cdot\mathbf{n}\,\mathrm{d}S \qquad (6.1.13)$$

where n is the outward normal to the boundary S of the region R.
   Equation (6.1.9) is now replaced by

$$T\iint_R \nabla w\cdot\nabla\phi\,\mathrm{d}x\,\mathrm{d}y - T\int_{S_1+S_2} w\frac{\partial\phi}{\partial n}\,\mathrm{d}S = \iint_R Qw\,\mathrm{d}x\,\mathrm{d}y. \qquad (6.1.14)$$

This is still an equation satisfied by the solution to the original differential equation and the condition on the boundary S_2, (6.1.3), can now be substituted to give

$$T\iint_R \nabla w\cdot\nabla\phi\,\mathrm{d}x\,\mathrm{d}y - T\int_{S_1} w\frac{\partial\phi}{\partial n}\,\mathrm{d}S = \iint_R Qw\,\mathrm{d}x\,\mathrm{d}y + \int_{S_2}\beta w\,\mathrm{d}S \qquad (6.1.15)$$

where the terms on the right-hand side are known. The source or sink term Q is represented in the same way as in section 5.2 using the Dirac delta generalized function. The Dirichlet known-head boundary condition S_1 has yet to be incorporated. Equation (6.1.15) is called a 'weak form' of the original differential equation. The requirements on the function $\phi(x, y)$ have been weakened because (6.1.15) now only includes the first and not the second derivatives of the function. It also conveniently incorporates the derivative boundary condition (6.1.3). It is now possible to use the piecewise linear approximation (6.1.5) constructed above.
   The standard Galerkin solution of the weak form (6.1.15) is obtained by taking

$$\phi(x, y) \approx \phi^h(x, y) = \sum_{j=1}^{J} \phi_j N_j(x, y)$$

and $w(x, y) = N_j(x, y)$, $j = 1, 2, \ldots, J$, in turn where J is the total number of nodes. (Readers who have used Fourier or Laplace transform methods will recognize the step just taken here as familiar: express the solution as a sum of terms in $\exp(i\omega x)$ or $\exp(-pt)$, multiply the differential equation by a typical term, and integrate by parts. This is another example of an integral transform.)
   Note that if eqn (6.1.1) is replaced by the time-dependent parabolic form

$$S\frac{\partial\phi}{\partial t} - T\nabla^2\phi = Q \qquad (6.1.16)$$

the time-derivative term is not involved in the integration by parts and (6.1.15) is replaced by

$$S \iint_R w \frac{\partial \phi}{\partial t} \, \mathrm{d}x \, \mathrm{d}y + T \iint_R \nabla w \cdot \nabla \phi \, \mathrm{d}x \, \mathrm{d}y - T \int_{S_1} w \frac{\partial \phi}{\partial n} \, \mathrm{d}S$$

$$= \iint_R Qw \, \mathrm{d}x \, \mathrm{d}y + \int_{S_2} \beta w \, \mathrm{d}S \quad (6.1.17)$$

where the storativity S is supposed constant.

The approximation (6.1.5) now becomes

$$\phi(x, y, t) \approx \sum_{j=1}^{J} \phi_j(t) N_j(x, y) \quad (6.1.18)$$

i.e. the approximate nodal values are now functions of time. Again the standard Galerkin method takes the functions $w(x, y)$ equal to the basis functions $N_j(x, y)$ in turn.

The result of the finite element approximation substituted in the Galerkin weak forms (6.1.15) is a system of simultaneous linear equations of the form

$$\mathbf{K}\boldsymbol{\phi} = \mathbf{F} \quad (6.1.19)$$

where $\boldsymbol{\phi}$ is the vector of nodal values. Eqn. (6.1.17) produces a system of equations of the form

$$\mathbf{M} \frac{\mathrm{d}}{\mathrm{d}t} \boldsymbol{\phi}(t) + \mathbf{K}\boldsymbol{\phi}(t) = \mathbf{F} \quad (6.1.20)$$

where $\boldsymbol{\phi}(t)$ is now the vector of nodal values. The matrixes $\mathbf{K}, \mathbf{M}$ involved in these equations have particular patterns which are discussed in the next section.

6.2 The finite element matrices

It is simpler here to deal with the more general time-dependent form (6.1.17) because it includes the two distinct types of finite element matrix.

Using the approximation (6.1.18) and $w = N_i(x, y)$, the first term on the left-hand side of (6.1.17) produces

$$S \sum_j \frac{\mathrm{d}}{\mathrm{d}t} \phi_j(t) \iint_R N_i(x, y) N_j(x, y) \, \mathrm{d}x \, \mathrm{d}y. \quad (6.2.1)$$

The matrix **M** whose (i, j)th entry is

$$m_{ij} = S \iint_R N_i(x, y) N_j(x, y) \, \mathrm{d}x \, \mathrm{d}y \qquad (6.2.2)$$

is called the mass matrix even when the physical problem being solved has no 'mass' involved. This is a legacy from the years of initial development of the FEM by structural engineers. The name 'mass matrix' can be regarded as a convenient label for the matrix whose entries involve the integral of the product of the basis functions undifferentiated as in (6.2.2).

The second term on the left of (6.1.17) produces

$$T \sum_j \phi_j(t) \iint_R \nabla N_i(x, y) \cdot \nabla N_j(x, y) \, \mathrm{d}x \, \mathrm{d}y. \qquad (6.2.3)$$

The matrix **K** whose (i, j)th entry is

$$k_{ij} = T \iint_R \nabla N_i(x, y) \cdot \nabla N_j(x, y) \, \mathrm{d}x \, \mathrm{d}y \qquad (6.2.4)$$

is called the stiffness matrix. Again this is a legacy from structural engineers and may just be taken as a convenient label for the matrix whose entries involve the inner product of the gradients of the basis functions. The mass and stiffness matrices can also be called the storage and conductance matrices in subsurface flow but the advantage of using the original labels is that they make it easier for the reader to move around the literature to pick up ideas from different physical contexts. The rule is not to focus on the label but on the underlying pattern of the mathematics.

Note that:

1. $m_{ij} = m_{ji}$, and $k_{ij} = k_{ji}$, i.e. the mass and stiffness matrices are symmetric.

2. The mass and stiffness matrix terms involve the storativity and the transmissivity which are internal properties of the system. (If the storativity and transmissivity are variables, $S(x, y)$, $T(x, y)$, they can be represented by average values over the element in each case.)

3. The terms on the right-hand side of (6.1.17) are known external effects which could vary for the same aquifer. They can be lumped

together as entries in a vector **F** (the external force vector) where

$$f_i = \iint_R QN_i(x, y)\, dx\, dy + \int_{S_2} \beta N_i(x, y)\, dS. \qquad (6.2.5)$$

4. The flow boundary condition on S_2 is included in eqn (6.1.17) so no more has to be done about this.

5. But the Dirichlet known-head (called 'essential') boundary conditions on S_1 still have to be included. In practice the finite element program assembles the mass and stiffness matrices (as described in the next section) first ignoring the essential boundary conditions. The values of $\partial\phi/\partial n$ on S_1 are not known and hence the equations of the system given by taking $w = N_i(x, y)$ where i is the number of a node on S_1 should be separated out and kept in reserve. There are then two ways of overwriting the known-head nodal values:

(a) The equations corresponding to nodes on S_1 are removed from the system and the known-head nodal values are included in the rest of the equations. The number of equations to solve is then equal to the number of unknown nodal values.

(b) Alternatively the equations corresponding to nodes i on S_1 are left in the system but overwritten so that the diagonal term on the left-hand side is multiplied by a large number E and the right-hand side is replaced by E times the correct value. The number of equations to be solved is then equal to the total number of nodes.

When all the nodal values are available the unknowns in the equations for nodes i on S_1 can be filled in to give the values of

$$T \int_{S_1} \frac{\partial \phi}{\partial n}\, N_i(x, y)\, dS \qquad (6.2.6)$$

and this gives the flow across the boundary in the neighbourhood of the known-head nodes (Lynch, 1984). Examples of the usefulness of this are given in Chapter 7.

The principal ingredients of the finite element program are:

(1) formation of element matrices;

(2) assembly into the global matrices;

(3) solution of the resultant system of algebraic equations.

The form of the element matrix is discussed in detail first for the simplest case of the linear triangle element. There are two ways of

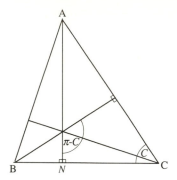

Figure 6.2.1: Triangle ABC.

expressing the element stiffness matrix for a linear triangle. The first way described here is NOT the way it is programmed but it is given here because it is useful to know how the element stiffness matrix depends on the angles of the triangle.

Figure 6.2.1 shows a general triangle ABC with corresponding basis functions $N_A(x, y)$, $N_B(x, y)$, $N_C(x, y)$. The linear triangle has just the three basis functions and the element stiffness matrix is of order 3. It is only necessary to get one diagonal term and one off-diagonal term to get the general pattern.

From (6.2.4) the entries in the stiffness matrix contain the integrals of the inner products of the gradients of the basis functions. Recall from section 6.1 that $N(x, y)$ is represented by a pyramid of height unity at A decreasing linearly to zero on BC. Thus the numerical value of the gradient of the part of $N_A(x, y)$ which influences triangle ABC (i.e. the slope of that face of the pyramid) is $(AN)^{-1}$ where N is the foot of the perpendicular from A to BC. Hence

$$|\nabla N_A(x, y)| = \frac{1}{AN} = \frac{1}{b \sin C} \tag{6.2.7}$$

with the usual notation (the sides of the triangle are a, b, c; the angle A is opposite side a, etc.). Of course since the basis functions are linear their gradients are constants. Thus the first diagonal term of the element stiffness matrix is

$$\iint_{\triangle ABC} |\nabla N_A(x, y)|^2 \, dx \, dy = \frac{ab \sin C}{2(b \sin C)^2} = \frac{a}{2b \sin C}$$

(using $\frac{1}{2}ab \sin C$ for the area of the triangle)

$$= \frac{(b \cos C + c \cos B)}{2b \sin C} = \tfrac{1}{2}(\cot C + \cot B). \qquad (6.2.8)$$

Also from Fig. 6.2.1

$$\iint_{\triangle ABC} \nabla N_A(x, y) \cdot \nabla N_B(x, y) \, dx \, dy$$

$$= \iint_{\triangle ABC} |\nabla N_A(x, y)||\nabla N_B(x, y)| \cos(\pi - C) \, dx \, dy$$

$$= -\frac{bc \sin A \cos C}{b \sin C \, c \sin A} = -\tfrac{1}{2} \cot C. \qquad (6.2.9)$$

Thus the element stiffness matrix $\mathbf{K}^{(e)}$ can be expressed entirely in terms of the cotangents of the angles:

$$\mathbf{K}^{(e)} = \frac{1}{2} \begin{bmatrix} \cot B + \cot C & -\cot C & -\cot B \\ \text{symmetric} & \cot C + \cot A & -\cot A \\ & & \cot A + \cot B \end{bmatrix}. \qquad (6.2.10)$$

This form of the element stiffness matrix shows up the effect of thin triangular elements because the numerical values of the cotangents increase rapidly as the angles get smaller or approach 180°. This leads to an ill-conditioned matrix $\mathbf{K}$ which should be avoided because then the computed solution of the system of equations (6.1.19) may bear no resemblance to the exact solution of this system (Conte and de Boor, 1980). There is also a lower limit on the time step in a conditionally stable time-stepping scheme (Wood, 1990). The form (6.2.10) also shows the effect of a right angle in a triangle element. Because $\cot 90° = 0$, if $C = 90°$ there is no term connecting nodes A and B in the element stiffness matrix.

Note that the element stiffness matrix must always be singular. This is evidently true for the matrix above because adding across the rows gives zeros. The reason why it is always true can be seen by considering the problem

$$\nabla^2 \phi = 0$$

in R with $\partial \phi / \partial n = 0$ all round the boundary of R, i.e. a vertically integrated aquifer with an impervious boundary so that there is no

flow and there is only an S_2-type boundary. The solution is that the
piezometric head ϕ can have any constant value; in the absence of any
further information the precise constant value is indeterminate. Now
suppose that the region R is covered by just one element. Equation
(6.2.9) gives the matrix of the set of equations to solve for the nodal
values. There are no fixed head node values to overwrite and any
constant value of $\phi = p$, say, will satisfy the equations:

$$(\cot B + \cot C)p - p \cot C - p \cot B = 0, \quad \text{etc.}$$

This is similarly true for any of the other elements introduced later.
It is often useful in checking a finite element program to verify that
the element stiffness matrix is singular.

In the programming the subroutines for computing the element
stiffness matrices use local coordinates called area coordinates. Figure
6.2.2 shows a general triangle with vertices labelled 1, 2, 3. A general
point P in this triangle has area coordinates given by

$$L_1 = \frac{\text{area of } \Delta P23}{\text{area of } \Delta 123}, \quad \text{etc.} \tag{6.2.11}$$

Each $L_j = 1$ at node j and is zero at the other nodes in the triangle
and linear in between; hence the basis functions $N_j(x, y)$ can be put
equal to $L_j, j = 1, 2, 3$. From (6.2.11) the area coordinates also satisfy

$$L_1 + L_2 + L_3 = 1.$$

The area of a triangle with vertices given by the global coordinates
(x_j, y_j), $j = 1, 2, 3$, is given by

$$\Delta_e = \frac{1}{2} \begin{vmatrix} 1 & x_1 & x_2 \\ 1 & x_2 & y_2 \\ 1 & x_3 & y_3 \end{vmatrix} \tag{6.2.12}$$

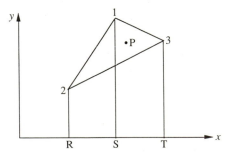

Figure 6.2.2: Area of triangle element.

with the vertices taken anticlockwise. (This can be verified by using the trapezium rule as in Fig. 6.2.2: area(123) = area(12RS) + area(1ST3) − area(2RT3)).) From the definition (6.2.11) the area coordinates are given by

$$L_1 = \frac{1}{2\Delta_e} \begin{vmatrix} 1 & x & y \\ 1 & x_2 & y_2 \\ 1 & x_3 & y_3 \end{vmatrix} \tag{6.2.13}$$

etc., where (x, y) is the position of the point P referred to global coordinates. That is,

$$L_1 = \frac{1}{2\Delta_e} (b_1 + c_1 x + d_1 y) \tag{6.2.14}$$

where

$$b_1 = \begin{vmatrix} x_2 & y_2 \\ x_3 & y_3 \end{vmatrix}, \quad c_1 = - \begin{vmatrix} 1 & y_2 \\ 1 & y_3 \end{vmatrix}, \quad d_1 = \begin{vmatrix} 1 & x_2 \\ 1 & x_3 \end{vmatrix}. \tag{6.2.15}$$

The global–local coordinate relationship is thus

$$x = L_1 x_1 + L_2 x_2 + L_3 x_3, \quad y = L_1 y_1 + L_2 y_2 + L_3 y_3. \tag{6.2.16}$$

Also from (6.2.14)

$$\frac{\partial L_1}{\partial x} = \frac{c_1}{2\Delta_e}, \quad \frac{\partial L_1}{\partial y} = \frac{d_1}{2\Delta_e} \tag{6.2.17}$$

etc., with cyclic rotation of subscripts.

Hence the element stiffness matrix has (i, j) entry

$$k_{ij}^{(e)} = \frac{1}{4\Delta_e^2} \iint_{\Delta_e} (c_i c_j + d_i d_j) \, dx \, dy$$

$$= \frac{1}{4\Delta_e} (c_i c_j + d_i d_j). \tag{6.2.18}$$

This is the form usually used in the program.

Assembly

The subroutine to calculate the element stiffness matrix is entered with the global coordinates of the three vertices of the triangle as required to calculate the entries $k_{ij}^{(e)}$ from (6.2.18). The program must specify, given the vertices, their local numbering 1, 2, 3. For example,

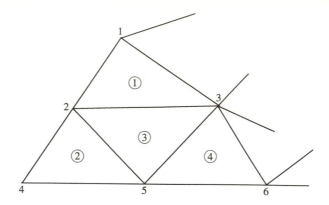

Figure 6.2.3: Local–global numbering.

node 1 has the largest x coordinate, if there are two such the one with the larger y coordinate etc. anticlockwise, and must include dual labelling of the nodes, e.g.:

NN(element number, local node number) = global node number.

Figure 6.2.3 shows an example with $NN(1, 1) = 3$, $NN(1, 2) = 1$, $NN(1, 3) = 2$, $NN(2, 1) = 5$, $NN(2, 2) = 2$, etc.

The element mass matrix is easily obtained using the useful formula

$$\iint_{\Delta_e} L_1^a L_2^b L_3^c \, dx \, dy = \frac{2a! \, b! \, c!}{(a + b + c + 2)!} \Delta_e. \tag{6.2.19}$$

Since

$$m_{ij}^{(e)} = \iint_{\Delta_e} L_i L_j \, dx \, dy$$

the element mass matrix is given by

$$\mathbf{M}^{(e)} = \frac{\Delta_e}{12} \begin{bmatrix} 2 & 1 & 1 \\ 1 & 2 & 1 \\ 1 & 1 & 2 \end{bmatrix}. \tag{6.2.20}$$

The contribution to the right-hand-side vector $\mathbf{F}$ in (6.1.20) may include terms of the form

$$\iint_{\Delta_e} L_1 f(x, y) \, dx \, dy \tag{6.2.21}$$

in the row whose number is the global number corresponding to the local number 1 here. Then $f(x, y)$ could be expressed in terms of L_1, L_2, L_3 using (6.2.16) and the integration performed using (6.2.19), but in a finite element program the integration is almost always performed numerically and this is discussed in section 6.6. The second term on the right of (6.1.15) only applies when $w(x, y) = N_i(x, y)$ where i is the number of a node on S_2. If β is a constant the contribution from an element here is

$$\beta \int_{S_2} N_i(x, y)\, \mathrm{d}S = \tfrac{1}{2}\beta \mathrm{AB}$$

where AB is the length of the side of the triangle along the boundary.

6.3 Some other types of element

Linear triangles are the easiest elements to use but in some problems quadrilateral elements are more suitable for covering the region. Also it may be worth while to use higher-degree polynomial approximations. In this section quadratic triangles and squares which are bilinear or include complete quadratic variation are introduced. In section 6.4 it is shown how these higher-degree approximations can be used for elements with curved sides which are especially useful in regions with curved boundaries.

The six-node quadratic triangle

Figure 6.3.1 shows the six-node triangle. The nodes are at the vertices and the midpoints of the sides. The area coordinates introduced in section 6.2 can again be used for the expression of the basis functions which are now quadratic.

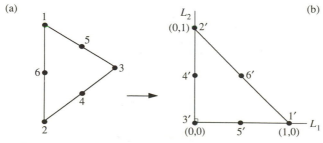

Figure 6.3.1: Six-node triangle, L_1, L_2 coordinates.

For the corner nodes:

$$N_1 = (2L_1 - 1)L_1, \quad \text{etc.} \tag{6.3.1}$$

satisfies the requirement that $N_1 = 1$ at node 1 and is zero at the other five nodes because $2L_1 = 1$ is the equation of the line joining nodes 5 and 6.

For the midside nodes:

$$N_4 = 4L_2L_3, \quad \text{etc.} \tag{6.3.2}$$

satisfies the requirements because $L_2 = L_3 = 1/2$ at node 4, L_2 is zero on the side 13, and L_3 is zero on the side 12 of the triangle.

For the derivatives in the integrand of the stiffness matrix terms

$$\frac{\partial N_1}{\partial x} = \frac{dN_1}{dL_1}\frac{\partial L_1}{\partial x} = (4L_1 - 1)\frac{c_1}{2\Delta_e} \tag{6.3.3}$$

from (6.2.17) etc.

$$\frac{\partial N_4}{\partial x} = 4L_2\frac{\partial L_3}{\partial x} + 4L_3\frac{\partial L_2}{\partial x} = \frac{2}{\Delta_e}(L_2c_3 + L_3c_2), \quad \text{etc.} \tag{6.3.4}$$

Because $L_1 + L_2 + L_3 = 1$ the basis functions can each be expressed in terms of L_1 and L_2 only. Thus the general triangle 123 in Fig. 6.3.1(a) can be mapped on to the right-angled triangle in Fig. 6.3.1(b). This is useful in connection with the numerical integration methods which are used in the program for the calculation of the entries in the element matrices.

Finite elements in two-dimensional problems are either triangles or quadrilaterals. Two types of square element are introduced next and general quadrilaterals are introduced in the next section.

The four-node bilinear square element

Figure 6.3.2. shows this element. The local coordinates are now referred to axes $0\xi, 0\eta$ with origin at the centre of the square whose vertices are $(-1, -1)$, $(1, -1)$, $(1, 1)$, and $(-1, 1)$, numbered 1, 2, 3, and 4 respectively as shown. The bilinear basis functions are

$$N_1 = \tfrac{1}{4}(1 - \xi)(1 - \eta), \quad \text{etc.} \tag{6.3.5}$$

These basis functions give linear terms plus a term in $\xi\eta$.

The eight-node square element

Figure 6.3.3 shows this element which has nodes at the corners and the midpoints of the sides. With the same local coordinate axes as

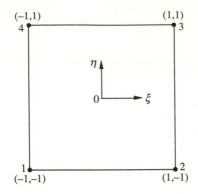

Figure 6.3.2: Four-node bilinear element.

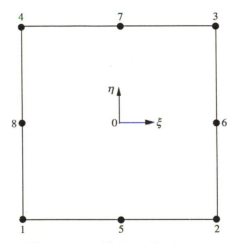

Figure 6.3.3: Eight-node element.

with the bilinear element and nodes numbered as in Fig. 6.3.3 the basis functions are now

corner nodes: $N_3 = \frac{1}{4}(1 + \xi)(1 + \eta)(\xi + \eta - 1)$,   etc.   (6.3.6)

midside nodes: $N_6 = \frac{1}{2}(1 - \eta^2)(1 + \xi)$,   etc.   (6.3.7)

Multiplying out these terms shows that they include a complete quadratic, i.e. terms in constant, ξ, η, ξ^2, $\xi\eta$, η^2 plus $\xi^2\eta$ and $\xi\eta^2$, eight degrees of freedom in all.

The nine-node square element

This element is like that shown in Fig. 6.3.3 with the addition of a node at the centre. The basis functions are

centre node: $N_0 = (1 - \xi^2)(1 - \eta^2)$ $\qquad\qquad\qquad$ (6.3.8)

corner node: $N_3 = \tfrac{1}{4}\xi\eta(1 + \xi)(1 + \eta)$ $\qquad\qquad$ (6.3.9)

midside node: $N_7 = \tfrac{1}{2}\eta(1 + \eta)(1 - \xi^2).$ $\qquad\qquad$ (6.3.10)

These basis functions include the terms as in the eight-node element plus $\xi^2\eta^2$, i.e. a biquadratic variation.

Three-dimensional elements

For three-dimensional groundwater flow problems a simple practical arrangement consists of dividing the region into horizontal layers filled with eight-node bricks and six-node triangular prisms adjacent to the surface as shown in Fig. 6.3.4. For an eight-node brick the local

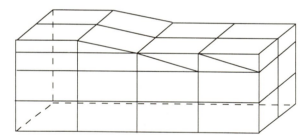

Figure 6.3.4: Bricks plus prisms.

coordinates are an extension of those for the four-node element, i.e. referred to axes 0ξ, 0η, 0ζ with origin at the centre. The basis functions are then

$$N_1 = \tfrac{1}{8}(1 - \xi)(1 - \eta)(1 - \zeta), \quad \text{etc.} \qquad (6.3.11)$$

For the six-node triangular prism a general point P can have coordinates (L_1, L_2, ζ), where L_1, L_2 are areal coordinates of P in the triangular section through P at right angles to the axis of the prism, and the ends of the prism are given by $\zeta = \pm 1$. The basis functions are then

$$N_1 = \tfrac{1}{2}L_1(1 - \zeta), \quad N_4 = \tfrac{1}{2}L_1(1 + \zeta), \quad \text{etc.} \qquad (6.3.12)$$

6.4 Isoparametric elements (Ergatoudis *et al.*, 1968)

Bilinear quadrilateral

With this ingenious idea the bilinear square becomes a general bilinear quadrilateral. The *same* basis functions are used for transforming the sides of the element as for the approximate solution of the differential equation:

$$x = \sum_1^4 x_j N_j(\xi, \eta), \qquad y = \sum_1^4 y_j N_j(\xi, \eta) \tag{6.4.1}$$

where the $N_j(\xi, \eta)$ are given in (6.3.5). Figure 6.4.1 shows the result. If the global coordinates of the vertices of the quadrilateral are (x_j, y_j), $j = 1, 2, 3, 4$, then any point P on the line 12 has global coordinates of the form $[\lambda x_1 + (1 - \lambda)x_2, \lambda y_1 + (1 - \lambda)y_2]$. From (6.4.1)

$$x_P = \lambda x_1 + (1 - \lambda)x_2, \qquad y_P = \lambda y_1 + (1 - \lambda)y_2 \tag{6.4.2}$$

correspond to a point P', in the ξ, η plane where

$$N_1(\xi, \eta) = \tfrac{1}{4}(1 - \xi)(1 - \eta) = \lambda, \qquad N_2(\xi, \eta) = \tfrac{1}{4}(1 + \xi)(1 - \eta) = (1 - \lambda).$$
$$\tag{6.4.3}$$

Solving for ξ, η gives $\xi = 1 - 2\lambda$, $\eta = -1$. Thus the isoparametric transformation sends the straight-line side 12 of the general quadrilateral into the side of the square $\eta = -1$ in the ξ, η plane and correspondingly with the other sides. The general quadrilateral in the global x, y coordinates is transformed into the square in ξ, η coordinates. The integrand for the numerical integration in computing the entries for the element matrices can be first formed in the context of the simple

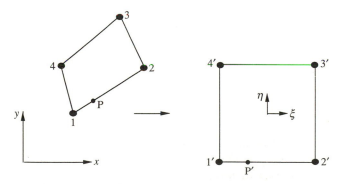

Figure 6.4.1: Four-node isoparametric element.

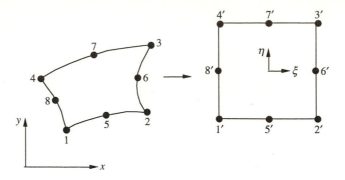

Figure 6.4.2: Isoparametric eight-node element.

square element and then transformed back into global coordinates to represent the effect of the general quadrilateral as described in section 6.7.

The eight-node isoparametric element

The same idea can be used with the eight-node element:

$$x = \sum_{1}^{8} x_j N_j(\xi, \eta), \qquad y = \sum_{1}^{8} y_j N_j(\xi, \eta) \qquad (6.4.4)$$

(Fig. 6.4.2). The element can now have curved sides represented by parabolas as the basis functions now contain complete quadratics. This is most useful in representing an aquifer with a curved boundary, then it would only be the side of the element on the boundary that would be curved; the others would be straight.

The six-node isoparametric triangle

Similarly this element employs

$$x = \sum_{1}^{6} x_j N_j(L_1, L_2), \qquad y = \sum_{1}^{6} y_j N_j(L_1, L_2) \qquad (6.4.5)$$

(Fig. 6.4.3) where the basis functions are given in section 6.3. This element is most useful where there is a curved boundary alongside which the elements can have one side curved; otherwise their sides are straight.

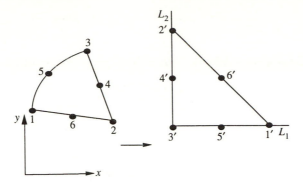

Figure 6.4.3: Isoparametric six-node triangle.

The use of isoparametric elements where the aquifer has a curved boundary gives a great improvement in accuracy. The result is 'more accurate and convenient than was ever achieved with finite differences' to quote Strang and Fix (1973).

Thacker (1980) gives a brief review of techniques for generating irregular computational grids suitable for use with the finite element method.

6.5 Accuracy, convergence, superconvergence

The accuracy of the numerical solution obtained by the finite element method is presented here as would be expected from the particular form of approximation used. With the linear triangle element, for instance, comparison with the Taylor series expansion (3.6.7) shows that if only linear terms are included in the approximation, there is an error proportional to the square of the 'mesh' and the second derivatives of the solution. The 'mesh' here is the measure of the size of the element; h, say, can be the length of the longest side. The size of the element, h, is essentially a ratio of the actual size to that of the region R, i.e. h is less than unity and the higher the power of h in the error the smaller the error. The error with linear triangles is expected to be proportional to

$$h^2 \text{ times the second derivatives of the exact solution.} \quad (6.5.1)$$

(Recall the image used in section 6.1 of the polyhedral surface of flat triangular plates approximating a curved surface.) Similarly the error in the gradient is expected to be proportional to

$$h \text{ times the second derivative of the exact solution.} \quad (6.5.2)$$

(For example, in one dimension Taylor's theorem with a remainder term gives

$$\phi(a + h) = \phi(a) + h\phi'(a) \quad \{\text{result of the piecewise linear}$$

approximation for the function}

$$+ \frac{h^2}{2} \phi''(a + \theta_1 h)$$

and

$$\phi'(a + h) = \phi'(a) \quad \{\text{result of the piecewise constant}$$

approximation for the gradient of the function}

$$+ h\phi''(a + \theta_2 h)$$

where $0 \leq \theta_1, \theta_2 \leq 1$.)

This is a local effect and if the solution has high second derivatives the error can be controlled by reducing the size of the elements in that locality.

Similarly with the six-node triangle element. The element basis functions contain a complete quadratic and continuing the Taylor expansion shows the error in the function now proportional to

$$h^3 \text{ times the third derivatives of the exact solution} \qquad (6.5.3)$$

and the error in the gradient of the function proportional to

$$h^2 \text{ times the third derivatives of the exact solution.} \qquad (6.5.4)$$

As the mesh size is decreased the approximate solution converges to the exact solution.

The four-node bilinear element contains the linear variation terms but not a complete quadratic. Hence the error with this is the same as in (6.5.1) and (6.5.2). The eight-node element contains a complete quadratic but not a complete cubic and the error is the same as in (6.5.3) and (6.5.4). The shape of the region often determines whether it is more convenient to use triangles or quadrilaterals.

Superconvergence of the gradient

Some elements have particular points where the gradient of the function as derived from the finite element solution has a higher-order accuracy than expected in general. This is called superconvergence (see Barlow, 1976; Zlamal, 1977).

Four-node quadrilaterals The centroid is the best point at which to evaluate the gradient. The error here is $O(h^2)$ when the elements are parallelograms.

Eight-node quadrilaterals The 2×2 Gauss points give higher-order accuracy for sampling the gradients. This has been proved (Lesaint and Zlamal, 1979) for elements close to parallelograms and is widely used in practice. This result conveniently fits in with using numerical integration at the same points.

Six-node triangles Levine (1985) has demonstrated superconvergence of the gradients at the midpoints of the sides provided the triangles can be paired to form parallelograms. The gradients can be recovered to greater accuracy by taking the average of the normal derivatives either side of a triangle edge and combining with the common gradient along the edge. Also the average of the gradients at the three edge midpoints gives a superconvergent value at the centroid. Hence it is better to program the numerical integration with the four-point formula using the midpoints of the sides described in section 6.6 so that the gradients of the basis functions are already available at these points.

Eight-node brick elements Barlow (1976) cites the centroid as the optimum point at which to evaluate the gradient.

These results are illustrated in some of the practical applications described in Chapter 7.

6.6 Numerical integration in the finite element method

In finite element programs the integration for the evaluation of the entries in the mass and stiffness matrices is usually done numerically. There are standard subroutines for triangles and quadrilaterals and a finite element package usually includes a selection. Hence in the finite element solution there are two sources of error:

(1) the error due to the approximation which has been discussed in section 6.5;

(2) the error due to the numerical integration.

It is essential to make sure that the error due to the numerical integration does not swamp the approximation error but it is also necessary not to waste CPU time by making the numerical integration more accurate than it need be. Strang and Fix (1973) give rules for the numerical integration of the stiffness matrix to be such that the error due to this is of the same order (i.e. same power of the element size h) as the finite element approximation error. The results of these rules are given here for the elements described in this chapter. There is no guidance given by Strang and Fix nor by Zienkiewicz and Taylor (1989) on the calculation of the mass matrix and the rules given here are based on the author's practical experience. The patch test described in section 6.7 can always be used in making a decision.

The basic idea of numerical integration is summarized for the two-dimensional case as

$$\iint\limits_{R} f(x, y)\, dx\, dy \approx \sum_{j=1}^{N} w_j f(x_j, y_j) \qquad (6.6.1)$$

where (x_j, y_j) are the points in the region R where the integrand $f(x, y)$ is 'sampled', w_j are the weights, and N is the number of points in the formula. There is a simplified version for the three-node linear triangle

The three-node linear triangle

With the linear triangle the gradients of the basis functions are constants, so the result is a constant times the area of the triangle. If the transmissivity in eqn (6.1.1) is $T(x, y)$ then the element stiffness matrix entries are of the form

$$\iint\limits_{\Delta_e} T(x, y) \nabla N_i \cdot \nabla N_j\, dx\, dy \qquad (6.6.2)$$

and this is approximated as

$$T(\bar{x}, \bar{y}) \iint\limits_{\Delta_e} \nabla N_i \cdot \nabla N_j\, dx\, dy \qquad (6.6.3)$$

where $(\bar{x}, \bar{y})$ is the centroid of the triangle, i.e. the transmissivity is sampled at the centroid. The entries for the element stiffness matrix are given in (6.2.18). If the transmissivity varies rapidly smaller elements should be used. The effect of this can be tested by experiment. The element mass matrix is given in (6.2.20). Again if the physical parameter multiplying the mass matrix term is variable it can be sampled at the centroid and if it varies rapidly smaller elements should be used.

The six-node triangle

The Strang and Fix rule for the stiffness matrix is that the numerical integration should be such that a quadratic is integrated exactly. There are two standard three-point formulae which will do this:

1. Sampling at the midpoints of the sides of the triangle whose areal coordinates are $(\frac{1}{2}, \frac{1}{2}, 0)$, $(\frac{1}{2}, 0, \frac{1}{2})$, $(0, \frac{1}{2}, \frac{1}{2})$ (Fig. 6.6.1(a)) with weights $\frac{1}{3}$. This is the better formula if the gradients are required (see section 6.5).

2. Sampling at the points half way between the vertices and the centroid, $(\frac{2}{3}, \frac{1}{6}, \frac{1}{6})$, $(\frac{1}{6}, \frac{2}{3}, \frac{1}{6})$, $(\frac{1}{6}, \frac{1}{6}, \frac{2}{3})$ (Fig. 6.6.1(b)) with weights $\frac{1}{3}$.

The mass matrix can be programmed using (6.3.1), (6.3.2), and (6.2.19).

The four-node square

For square and quadrilateral elements the numerical integration rules are extensions of the one-dimension Gauss integration rules given in books on elementary numerical analysis such as Conte and de Boor (1980). For the four-node square the Strang and Fix rules suggest that one-point Gauss is sufficient. This is exact for a linear integrand $f(\xi, \eta) = 1 + \xi + \eta$,

$$\int_{-1}^{1} \int_{-1}^{1} (1 + \xi + \eta) \, d\xi \, d\eta = f(0, 0) \text{ times the area.}$$

But this may not be suitable because the element stiffness matrix given by this one-point sampling is

$$\frac{1}{2} \begin{bmatrix} 1 & 0 & -1 & 0 \\ 0 & 1 & 0 & -1 \\ -1 & 0 & 1 & 0 \\ 0 & -1 & 0 & 1 \end{bmatrix} \tag{6.6.4}$$

(a)

(b)

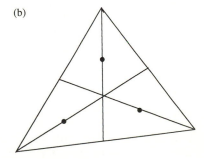

Figure 6.6.1: (a, b) Sampling points for a triangle.

when it ought to be

$$\frac{1}{6}\begin{bmatrix} 4 & -1 & -2 & -1 \\ -1 & 4 & -1 & -2 \\ -2 & -1 & 4 & -1 \\ -1 & -2 & -1 & 4 \end{bmatrix}. \qquad (6.6.5)$$

The correct element stiffness matrix (6.6.5) is singular (as it always should be); it has only one zero eigenvalue corresponding to the eigenvector

$$[1, 1, 1, 1]^{\mathrm{T}}. \qquad (6.6.6)$$

(In another legacy from structural engineers this is called the 'rigid-body mode'.) This corresponds to the arbitrary constant solution described in section 6.2 when there is no flow on the boundary. The one-point sampling matrix (6.6.4) has two zero eigenvalues, one corresponding to the eigenvector (6.6.6) and one corresponding to a spurious mode

$$[1, -1, 1, -1]^{\mathrm{T}}. \qquad (6.6.7)$$

This can produce a high-frequency oscillation across the elements like a choppy sea; this is an example of 'noise'. It is particularly likely to occur when there are not many Dirichlet fixed head nodes on the boundary of the aquifer. Hence it is safer to use the four-point Gauss formula which for the standard square element with local coordinates (ξ, η), $-1 \leq \xi, \eta < 1$, samples at the points

$$\left(\pm\frac{1}{\sqrt{3}}, \pm\frac{1}{\sqrt{3}} \right). \qquad (6.6.8)$$

(Note: Spurious oscillations should always be investigated and not simply averaged out. They may be caused, for example, by under-integration or elements too large. See Appendix 3.)

Experiments made with a time-dependent problem and with parallelogram elements (described in Chapter 7) show that the four-point Gauss formula is sufficient for the mass matrix also.

The eight-node square

The Strang and Fix rule for the stiffness matrix is to use the four-point Gauss formula as above. For the mass matrix it may again be necessary to make experiments using the patch test described in section 6.7.

Isoparametric elements

The element stiffness matrix requires entries of the form

$$\iint_{\text{el}} \left(\frac{\partial N_i}{\partial x} \frac{\partial N_j}{\partial x} + \frac{\partial N_i}{\partial y} \frac{\partial N_j}{\partial y} \right) \mathrm{d}x \, \mathrm{d}y \qquad (6.6.9)$$

where x, y are the global coordinates and el stands for the element. It is necessary to transform the integrand to (ξ, η) coordinates so as to use the standard square element basis functions and numerical integration formulae. The isoparametric equations (6.4.1) give x, y in terms of ξ, η but in general they cannot be inverted to give ξ, η in terms of x, y. Hence the chain rule (Spencer *et al.*, 1977) has to be used:

$$\frac{\partial N_i}{\partial \xi} = \frac{\partial N_i}{\partial x} \frac{\partial x}{\partial \xi} + \frac{\partial N_i}{\partial y} \frac{\partial y}{\partial \xi} \qquad (6.6.10)$$

$$\frac{\partial N_i}{\partial \eta} = \frac{\partial N_i}{\partial x} \frac{\partial x}{\partial \eta} + \frac{\partial N_i}{\partial y} \frac{\partial y}{\partial \eta} \qquad (6.6.11)$$

i.e.

$$\begin{bmatrix} \dfrac{\partial N_i}{\partial \xi} \\[2mm] \dfrac{\partial N_i}{\partial \eta} \end{bmatrix} = \mathbf{J} \begin{bmatrix} \dfrac{\partial N_i}{\partial x} \\[2mm] \dfrac{\partial N_i}{\partial y} \end{bmatrix}$$

where $\mathbf{J}$ is the Jacobian matrix of the transformation given by

$$\mathbf{J} = \begin{bmatrix} \dfrac{\partial x}{\partial \xi} & \dfrac{\partial y}{\partial \xi} \\[3mm] \dfrac{\partial x}{\partial \eta} & \dfrac{\partial y}{\partial \eta} \end{bmatrix} \qquad (6.6.12)$$

and which can be obtained from (6.4.1). Hence the derivatives required for the integrand in (6.6.9) can be obtained from

$$\begin{bmatrix} \dfrac{\partial N_i}{\partial x} \\[2mm] \dfrac{\partial N_i}{\partial y} \end{bmatrix} = \mathbf{J}^{-1} \begin{bmatrix} \dfrac{\partial N_i}{\partial \xi} \\[2mm] \dfrac{\partial N_i}{\partial \eta} \end{bmatrix} \qquad (6.6.13)$$

and

$$= \frac{1}{|\mathbf{J}|} \begin{bmatrix} \dfrac{\partial y}{\partial \eta} & -\dfrac{\partial y}{\partial \xi} \\[4mm] -\dfrac{\partial x}{\partial \eta} & \dfrac{\partial x}{\partial \xi} \end{bmatrix} \begin{bmatrix} \dfrac{\partial N_i}{\partial \xi} \\[4mm] \dfrac{\partial N_i}{\partial \eta} \end{bmatrix} \qquad (6.6.14)$$

where $|\mathbf{J}|$ is the determinant of the Jacobian matrix (Spencer *et al.*, 1977), and in the usual notation

$$|\mathbf{J}| = \frac{\partial(x, y)}{\partial(\xi, \eta)}. \qquad (6.6.15)$$

Hence

$$\frac{\partial N_i}{\partial x} = \frac{1}{|\mathbf{J}|} \left[\frac{\partial N_i}{\partial \xi} \frac{\partial y}{\partial \eta} - \frac{\partial N_i}{\partial \eta} \frac{\partial y}{\partial \xi} \right] \qquad (6.6.16)$$

and

$$\frac{\partial N_i}{\partial y} = \frac{1}{|\mathbf{J}|} \left[-\frac{\partial N_i}{\partial \xi} \frac{\partial x}{\partial \eta} + \frac{\partial N_i}{\partial \eta} \frac{\partial x}{\partial \xi} \right] \qquad (6.6.17)$$

where all the right-hand-side terms in (6.6.16) and (6.6.17) are obtainable from the isoparametric transformation. It is not necessary to go into details here to see the result is that

$$\iint_{\text{el}} \nabla N_i \cdot \nabla N_j \, \mathrm{d}x \, \mathrm{d}y = \iint_{\text{el}} \frac{1}{|\mathbf{J}|^2} (\text{polynomial in } \xi, \eta) |\mathbf{J}| \, \mathrm{d}\xi \, \mathrm{d}\eta$$

$$= \iint_{\text{el}} \frac{1}{|\mathbf{J}|} (\text{polynomial in } \xi, \eta) \, \mathrm{d}\xi \, \mathrm{d}\eta$$

and in general the integrand here is a rational function (a ratio of two polynomials) because $|\mathbf{J}|$ is itself an algebraic function of ξ, η. The Gauss numerical integration rules are exact for polynomials but also quite good for rational functions away from singularities, thus it is important not to have $|\mathbf{J}|$ near zero. $|\mathbf{J}|$ is a constant for a parallelogram so that if the elements do not deviate too far from parallelograms the four-point Gauss rule should be safe for four-node or eight-node quadrilaterals. $|\mathbf{J}| = 0$ implies a functional relationship between

$$x = \sum x_j N_j(\xi, \eta) \quad \text{and} \quad y = \sum y_j N_j(\xi, \eta)$$

which happens if the quadrilateral collapses into a triangle *and* $|\mathbf{J}|$

is evaluated on this line. The latter does not happen if interior Gauss points are used for the numerical integration. The quadrilateral elements should not in any case have internal angles greater than 180° because then the determinant $|\mathbf{J}|$ changes sign.

The general rule for the numerical integration in the finite element method is to have various subroutines available and to use the patch test described in the next section.

6.7 The patch test

The patch test is used to test the consistency of the formulation of the problem, the program, the adequacy of the numerical integration, and anything else with which it is desired to experiment. Suppose data are inserted on the boundary of a patch of elements such that the exact solution is a polynomial of degree r and the basis functions are such that all polynomials of degree r are included. Then the program should give the exact solution and it is then said to have 'passed the patch test'. It is convenient to use a square patch but the elements must be typical. Figures 6.7.1(a) and (b) show examples with general triangles and general quadrilaterals.

Figures 6.7.2–7 (from Griffiths, 1978) show a typical set of patch tests designed to test eight-node elements for use with a large aquifer model. In each case the exact solution is a quadratic polynomial which satisfies

$$\nabla^2 \phi = 0 \qquad (6.7.1)$$

and the exact values are substituted at the boundary nodes. Figure 6.7.2 shows that with parallelogram elements the exact solution is obtained at the interior nodes. Figure 6.7.3 shows small errors resulting from the deviation from the parallelogram measured by the variation of $|\mathbf{J}|$; using 2×2 or 3×3 Gauss points makes little difference.

(a) (b)

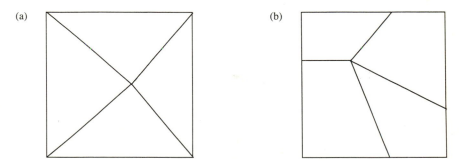

Figure 6.7.1: (a, b) Patch test, triangles and quadrilaterals.

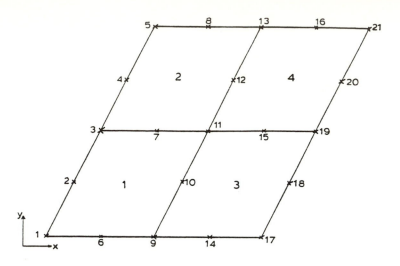

Node	Coordinates	True Solution	F.E. Solution
1	(0,0)	12.0	
2	(0.5,1)	13.25	
3	(1,2)	14.0	
4	(1.5,3)	14.25	
5	(2,4)	14.0	
6	(1,0)	14.0	
7	(2,2)	20.0	20.0
8	(3,4)	24.0	
9	(2,0)	18.0	
10	(2.5,1)	23.25	23.25
11	(3,2)	28.0	28.0
12	(3.5,3)	32.25	32.25
13	(4,4)	36.0	
14	(3,0)	24.0	
15	(4,2)	38.0	38.0
16	(5,4)	50.0	
17	(4,0)	32.0	
18	(4.5,1)	41.25	
19	(5,2)	50.0	
20	(5.5,3)	58.25	
21	(6,4)	66.0	

Element	Variation in $\lvert J \rvert$ $(\min \lvert J \rvert / \max \lvert J \rvert)$
1	1.0
2	1.0
3	1.0
4	1.0

Figure 6.7.2: Patch test results.

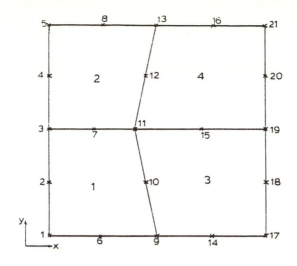

Node	Coord-inates	True Solution	F.E. Solution		Percentage Error	
			2x2 Gauss	3x3 Gauss	2x2 Gauss	3x3 Gauss
1	(0,0)	12.0				
2	(0,1)	12.0				
3	(0,2)	10.0				
4	(0,3)	6.0				
5	(0,4)	0.0				
6	(1,0)	14.0				
7	(0.8,2)	13.04	13.044438	13.044144	+0.034	+0.032
8	(1,4)	6.0				
9	(2,0)	18.0				
10	(1.8,1)	18.84	18.844699	18.844482	+0.025	+0.024
11	(1.6,2)	17.36	17.350562	17.351012	−0.054	−0.052
12	(1.8,3)	16.44	16.444699	16.444482	+0.029	+0.027
13	(2,4)	14.0				
14	(3,0)	24.0				
15	(2.8,2)	26.24	26.244968	26.244830	+0.019	+0.018
16	(3,4)	24.0				
17	(4,0)	32.0				
18	(4,1)	36.0				
19	(4,2)	38.0				
20	(4,3)	38.0				
21	(4,4)	36.0				

Element	Variation in $|J|$ $(\min|J|/\max|J|)$	
	2x2 Gauss	3x3 Gauss
1	0.87943	0.84150
2	0.87943	0.84150
3	0.90026	0.86842
4	0.90026	0.86842

Figure 6.7.3: Patch test results.

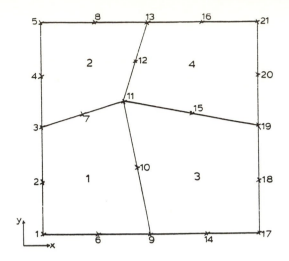

Node	Coordinates	True Solution	F.E. Solution		
			2x2	3x3	4x4
1	$(0,0)$	16.0			
2	$(0,1)$	15.0			
3	$(0,2)$	12.0			
4	$(0,3)$	7.0			
5	$(0,4)$	10.0			
6	$(1,0)$	17.0			
7	$(\frac{3}{4},2\frac{1}{4})$	11.5	11.5	11.5	11.5
8	$(1,4)$	10.0			
9	$(2,0)$	20.0			
10	$(1\frac{3}{4},1\frac{1}{4})$	17.5	17.5	17.5	17.5
11	$(1\frac{3}{4},2\frac{3}{4})$	12.0	12.0	12.0	12.0
12	$(1\frac{3}{4},3\frac{1}{4})$	8.5	8.5	8.5	8.5
13	$(2,4)$	4.0			
14	$(3,0)$	25.0			
15	$(2\frac{3}{4},2\frac{1}{4})$	18.5	18.5	18.5	18.5
16	$(3,4)$	9.0			
17	$(4,0)$	32.0			
18	$(4,1)$	31.0			
19	$(4,2)$	28.0			
20	$(4,3)$	23.0			
21	$(4,4)$	16.0			

Element	Variation in $|J|$ $(\min|J|/\max|J|)$		
	2x2 Gauss	3x3 Gauss	4x4 Gauss
1	0.74773	0.67553	0.64570
2	0.67721	0.58957	0.55394
3	0.79296	0.73172	0.70615
4	0.74773	0.67553	0.64570

Figure 6.7.4: Patch test results.

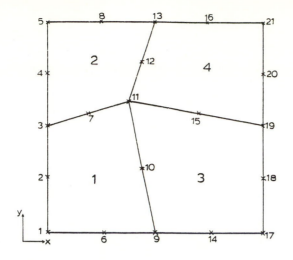

Node	Coord-inates	True S-olution	F.E. Solution			Percentage Error		
			2x2	3x3	4x4	2x2	3x3	4x4
1	$(0,0)$	16.0						
2	$(0,1)$	15.0						
3	$(0,2)$	12.0						
4	$(0,3)$	7.0						
5	$(0,4)$	0.0						
6	$(1,0)$	20.0						
7	$(\frac{3}{4},2\frac{1}{4})$	20.5	20.472868	20.475082	20.475102	-0.132	-0.122	-0.121
8	$(1,4)$	20.0						
9	$(2,0)$	26.0						
10	$(1\frac{1}{2},1\frac{1}{4})$	31.5	31.468506	31.469207	31.469204	-0.099	-0.098	-0.097
11	$(1\frac{3}{4},2\frac{3}{4})$	31.5	31.559274	31.555995	31.555974	$+0.188$	$+0.178$	$+0.178$
12	$(1\frac{3}{4},3\frac{1}{4})$	36.5	36.472868	36.475082	36.475102	-0.074	-0.068	-0.068
13	$(2,4)$	42.0						
14	$(3,0)$	34.0						
15	$(2\frac{3}{4},2\frac{1}{4})$	51.5	51.468506	51.469207	51.469204	-0.061	-0.060	-0.060
16	$(3,4)$	66.0						
17	$(4,0)$	44.0						
18	$(4,1)$	59.0						
19	$(4,2)$	72.0						
20	$(4,3)$	83.0						
21	$(4,4)$	92.0						

Element	Variation in $|J|$ $(\min|J|/\max|J|)$		
	2x2	3x3	4x4
1	0.74773	0.67553	0.64570
2	0.67721	0.58957	0.55394
3	0.79296	0.73172	0.70615
4	0.74773	0.67553	0.64570

Figure 6.7.5: Patch test results.

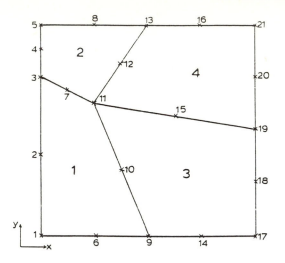

Node	Coord- inates	True S- olution	F.E. Solution			Percentage Error		
			2x2 Gauss	3x3 Gauss	4x4 Gauss	2x2	3x3	4x4
1	$(0,0)$	16.0						
2	$(0,1\frac{1}{2})$	13.75						
3	$(0,3)$	7.0						
4	$(0,3\frac{1}{2})$	3.75						
5	$(0,4)$	0.0						
6	$(1,0)$	17.0						
7	$(\frac{1}{2},2\frac{3}{4})$	8.6875	8.7056443	8.7035620	8.7035063	+ 0.209	+ 0.185	+ 0.184
8	$(1,4)$	1.0						
9	$(2,0)$	20.0						
10	$(1\frac{1}{2},1\frac{1}{4})$	16.6875	16.708627	16.706085	16.705973	+ 0.127	+ 0.111	+ 0.110
11	$(1,2\frac{1}{2})$	10.75	10.704931	10.708424	10.708555	− 0.419	− 0.387	− 0.386
12	$(1\frac{1}{2},3\frac{1}{4})$	7.6875	7.7107825	7.7107693	7.7107250	+ 0.303	+ 0.302	+ 0.302
13	$(2,4)$	4.0						
14	$(3,0)$	25.0						
15	$(2\frac{1}{2},2\frac{1}{4})$	17.1875	17.212439	17.212332	17.212350	+ 0.145	+ 0.145	+ 0.144
16	$(3,4)$	9.0						
17	$(4,0)$	32.0						
18	$(4,1)$	31.0						
19	$(4,2)$	28.0						
20	$(4,3)$	23.0						
21	$(4,4)$	16.0						

Element	Variation in $|J|$ (min $|J|$/max $|J|$)		
	2x2 Gauss	3x3 Gauss	4x4 Gauss
1	0.55198	0.44165	0.39804
2	0.55198	0.44165	0.39804
3	0.72792	0.65118	0.61962
4	0.67721	0.58957	0.55394

Figure 6.7.6: Patch test results.

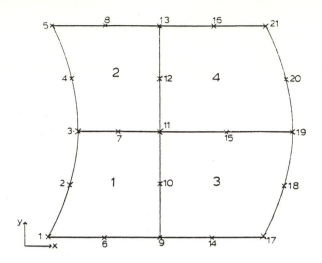

Node	Coord-inates	True Solution	F.E. Solution			Percentage Error		
			2x2 Gauss	3x3 Gauss	4x4 Gauss	2x2	3x3	4x4
1	(0,0)	16.0						
2	(0.375,1)	15.140625						
3	(Q5,2)	12.25						
4	(0.375,3)	7.140625						
5	(0.4)	0.0						
6	(1,0)	17.0						
7	(1.25,2)	13.5625	13.675227	13.673691	13.673668	+ 0.83	0.82	− 0.82
8	(1,4)	1.0						
9	(2,0)	20.0						
10	(2,1)	19.0	18.988273	18.990304	18.990424	− 0.062	− 0.051	− 0.050
11	(2,2)	16.0	16.028234	16.028533	16.028419	+ 0.177	+ 0.178	+ 0.178
12	(2,3)	11.0	10.988273	10.990304	10.990474	− 0.107	− 0.088	− 0.087
13	(2,4)	4.0						
14	(3,0)	25.0						
15	(3.25,2)	22.5625	22.475337	22.482969	22.482954	− 0.386	− 0.353	− 0.353
16	(3,4)	9.0						
17	(4,0)	32.0						
18	(4.375,1)	34.140625						
19	(4.5,2)	32.25						
20	(4.375,3)	26.140625						
21	(4,4)	16.0						

Element	Variation in J (min J /max J)		
	2x2 Gauss	3x3 Gauss	4x4 Gauss
1	0.84059	0.79547	0.77725
2	0.84059	0.79547	0.77725
3	0.88348	0.84468	0.82760
4	0.88348	0.84468	0.82760

Figure 6.7.7: Patch test results.

Figure 6.7.4 shows how important it is not to give an atypically accurate solution by introducing symmetry into the patch test. The exact solution used is

$$\phi = 16 + x^2 - y^2 \qquad (6.7.2)$$

and the finite element mesh is symmetric about the line through the nodes 5, 11, 17. The finite element solutions are all exact. Figure 6.7.5 shows the errors resulting from using another symmetric mesh but with a more general polynomial

$$\phi = 16 + 3x + x^2 + 4xy - y^2. \qquad (6.7.3)$$

Figure 6.7.6 shows the results of further distortion of the elements and Fig. 6.7.7 the results of using elements with curved sides.

   The results of these patch tests are used to produce the graph in Fig. 6.7.8. This is used to determine how the distortion of the element away from a parallelogram, as measured by the variation in the Jacobian $|\mathbf{J}|$, should be controlled in order to keep the error in the solution within acceptable limits. The conclusions are that:

(1) to limit the error due to this cause to a maximum of 0.05 at any node the limiting values of the ratio

$$\frac{\min|\mathbf{J}|}{\max|\mathbf{J}|}$$

should be taken as 0.55, 0.425, 0.40 for 2×2, 3×3 and 4×4 Gauss rules respectively.

(2) The extra CPU time required for the 3×3 and 4×4 Gauss rules is not justified and the 2×2 Gauss rule is used for the rest of the project. This demonstrates the usefulness of the patch test.

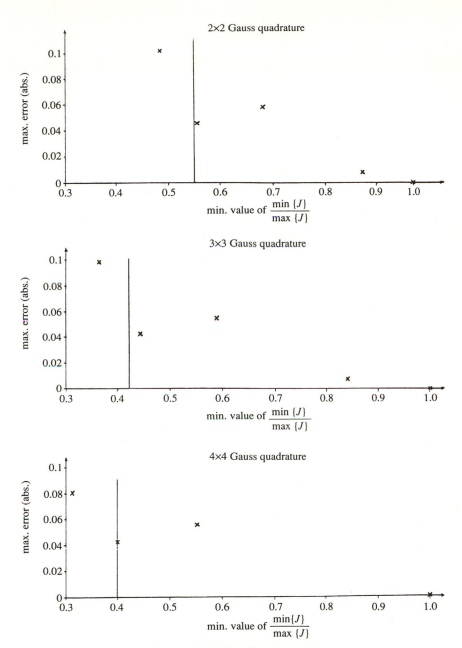

Figure 6.7.8: Results of patch tests.

7 Groundwater flow problems solved by finite elements

7.1 Introduction

The basic formulation of the groundwater flow equation is described in section 1.7. Appendix 1 shows how the three-dimensional equation can be vertically integrated to give a two-dimensions-in-plan equation suitable for aquifer modelling. This equation is non-linear but further simplifications can be made to make it linear. This chapter gives examples of the solution of groundwater flow problems by finite element methods. There is a general discussion of the solution of the non-linear aquifer equation (A1.18). The linear form using estimated transmissivities can be regarded as a particular case of this.

For the three-dimensional case it is possible to solve for the groundwater flow up to the phreatic surface but this is difficult because the position of this surface is initially unknown. This problem with some simplifying assumptions is discussed in section 7.6. Section 7.7 deals with the solution of the three-dimensional non-linear Richards' equation representing groundwater flow in the saturated and un-saturated regions up to the surface of the ground. This can be combined with surface flow with allowance for:

(1) how much of the effective precipitation penetrates the ground depending on the state of the ground just below the surface; and

(2) when the ground next to the surface is saturated the water may emerge. This includes the seepage face effect.

With the groundwater flow examples here the space discretization is done by finite element methods and the time-stepping scheme is the simple one-step method introduced in Chapter 3. The surface flow is modelled by using finite difference methods only because this is simplest. This chapter also includes discussion of:

(1) the question of obtaining the flow across the known-head boundaries in order to check on the mass balance;

(2) the question whether lumped or distributed mass matrices should be used for the modelling of the groundwater flow; and

(3) suitable initial conditions for the time-dependent problems.

7.2 Aquifer models—general considerations

As shown in Appendix 1 the two-dimensions-in-plan equation for aquifer modelling is given by

$$S\frac{\partial h}{\partial t} = \frac{\partial}{\partial x}\left(\bar{K}_x(h - h_b)\frac{\partial h}{\partial x}\right) + \frac{\partial}{\partial y}\left(\bar{K}_y(h - h_b)\frac{\partial h}{\partial y}\right) + Q \qquad (7.2.1)$$

where S is the effective storativity, h is the height of the water table in a phreatic (unconfined) aquifer, h_b is the height of the bottom of the aquifer, Q is the source or sink term, and $\bar{K}_x$, $\bar{K}_y$ are the vertically averaged hydraulic conductivities.

As remarked in section 1.7 it is often considered sufficient to solve the steady-state problem

$$-\frac{\partial}{\partial x}\left(\bar{K}_x(h - h_b)\frac{\partial h}{\partial x}\right) - \frac{\partial}{\partial y}\left(\bar{K}_y(h - h_b)\frac{\partial h}{\partial y}\right) = Q. \qquad (7.2.2)$$

The following discussion is simplified by supposing that the ground is isotropic, i.e.

$$\bar{K}_x = \bar{K}_y = \bar{K}. \qquad (7.2.3)$$

Equation (7.2.2) has the same form as eqn (6.1.1) used to introduce the finite element method in section 6.1 except for the extra terms $\bar{K}(h - h_b)$ which have to be included under the integral sign in the integration by parts so that eqn (6.1.15) is replaced by

$$\iint_R \bar{K}(h - h_b)\nabla w \cdot \nabla \phi \, dx \, dy - \int_{S_1} \bar{K}(h - h_b)w\frac{\partial \phi}{\partial n} \, dS$$

$$= \iint_R Qw \, dx \, dy + \int_{S_2} \beta w \, dS. \qquad (7.2.4)$$

Recall from Chapter 6 that for the standard finite element solution the test functions w in (7.2.4) are replaced by the basis functions $N_j(x, y)$ in turn. When j is the number of a boundary node there must be a boundary condition attached which is either 'known head' when node j is on S_1 or 'known flow' when node j is on S_2. The boundary node equations are treated in a similar way to that described in section 6.2:

1. When $w = N_j(x, y)$ and node j is on S_2 then the integral over S_1 in (7.2.4) is zero because w is zero here.

2. When $w = N_j(x, y)$ and node j is on S_1 the equation as it stands cannot be used because the flow term

$$\bar{K}(h - h_b) \frac{\partial h}{\partial n}$$

is not known. Since S_1 is the 'known-head' boundary the value of h_j is known and for the equations of this type:

(a) Either these equations are removed from the system (but preserved because after the rest of the equations are solved the other nodal values of h can be substituted to retrieve the flow terms (Lynch, 1984)). The known values of h_j are substituted in the rest of the equations. The number of simultaneous equations to be solved is then equal to the number of unknown nodal values.

(b) Alternatively the equations with $w = N_j(x, y)$ where node j is on S_1 are copied into a reserve (so that again the flow integral can be recovered) and then in the original system overwritten so that each has a large number times the h_j on the right-hand side and the correct value on the left-hand side. The number of simultaneous equations to be solved is then equal to the total number of nodes.

The stiffness matrix entries from (7.2.4) are now dependent on the unknown h:

$$k_{ij}(h) = \iint_R \bar{K}(h - h_b) \nabla N_i(x, y) \cdot \nabla N_j(x, y) \, dx \, dy. \qquad (7.2.5)$$

Thus there is a non-linear system of equations to be solved:

$$\mathbf{K}(h)\mathbf{h} = \mathbf{F}. \qquad (7.2.6)$$

The non-linear problem here can be solved by an iterative method (see Appendix 2) and at any stage the matrix used for solving for the $(k + 1)$th iterate, h^{k+1}, is formed by using the solution h^k from the previous iteration.

Jacoby (1982) in the Lower Colne example described in the next section takes

$$\mathbf{K}(h^k)\mathbf{h}^{k+1} = \mathbf{F} \qquad (7.2.7)$$

accelerates the convergence of the iteration by taking

$$\hat{\mathbf{h}}^{k+1} = \omega \mathbf{h}^{k+1} + (1 - \omega)\hat{\mathbf{h}}^k \qquad (7.2.8)$$

and reports that $\omega \approx 0.85$ gives most rapid convergence. The convergence criterion is taken as

$$\left| \frac{h_j^{k+1} - h_j^k}{h_j^k} \right| < \varepsilon, \quad j = 1, 2, \ldots, N \qquad (7.2.9)$$

with $\varepsilon = 10^{-3}$ regarded as satisfactory (h_j^k is not allowed to be zero).

The program used for the coastal aquifer described in section 7.3 uses a modified version of the 'dog-leg' algorithm of Powell (1970). This gives rapid convergence with, on average, only three iterations for a model with 783 nodes.

For the linearized version the term $\bar{K}(h - h_b)$ is replaced by estimated values of the transmissivity $T(x, y)$ so that the stiffness matrix entries can now be represented by

$$k_{ij} = \bar{T} \iint\limits_{R} \nabla N_i(x, y) \cdot \nabla N_j(x, y) \, \mathrm{d}x \, \mathrm{d}y \qquad (7.2.10)$$

where $\bar{T}$ is the average value of the transmissivity over the element. If the transmissivity varies considerably in part of the region the elements should be smaller here. The values of the transmissivity are obtained from site pumping tests. Figure 1.7.2 (from Connorton, 1980) shows the areal variation of transmissivity for the Lambourn aquifer.

As shown in section 6.2 the result of the finite element discretization of eqn (7.2.1) is a system of ordinary differential equations. With the non-linear stiffness matrix whose entries are given in eqn (7.2.5) this system of ordinary differential equations can be written as

$$\mathbf{M} \frac{\mathrm{d}\mathbf{h}}{\mathrm{d}t} + \mathbf{K}(h)\mathbf{h} = \mathbf{F} \qquad (7.2.11)$$

where $\mathbf{M}$ is the mass matrix whose terms are given by

$$m_{ij} = \bar{S} \iint\limits_{R} N_i(x, y)N_j(x, y) \, \mathrm{d}x \, \mathrm{d}y. \qquad (7.2.12)$$

Here $\bar{S}$ is the local average value of the effective storativity, the vector $\mathbf{h}$ is the vector of nodal values of the head, and the right-hand-side vector $\mathbf{F}$ contains the known-head values as well as any source or sink terms. The system of ordinary differential equations (7.2.11) as explained in Chapter 3 can be solved by the implicit scheme

$$\mathbf{M} \frac{(\mathbf{h}_{n+1} - \mathbf{h}_n)}{\Delta t} + \mathbf{K}(h_{n+1})\mathbf{h}_{n+1} = \bar{\mathbf{F}} \qquad (7.2.13)$$

where Δt is the time step, $\mathbf{h}_n$ is the approximate value of the vector of nodal values at time $n\Delta t$, and $\bar{\mathbf{F}}$ is an average value of the vector $\mathbf{F}$ over the time step. This non-linear equation is solved iteratively by taking

$$[\mathbf{M} + \Delta t \mathbf{K}(h_{n+1}^k)]\mathbf{h}_{n+1}^{k+1} = \mathbf{M}\mathbf{h}_n + \Delta t\bar{\mathbf{F}} \qquad (7.2.14)$$

$k = 0, 1, 2, \ldots$ until convergence, using $\mathbf{h}_{n+1}^0 = \mathbf{h}_n$, the converged solution from the previous time step as the starting value for the iteration. Provided the time steps are reasonably small, this should give rapid convergence.

7.3 Aquifer models—general procedure

Choice of boundaries

As remarked in section 1.7 it is often necessary to model a larger area in order to have suitable boundaries. For example, Fig. 7.3.1 (from Connorton and van Beeston, 1983) shows the area of the Lower Colne Gravels aquifer in which detailed results are required. In order to have convenient boundary conditions the total area has to be increased so as to be bounded by

(1) the River Thames to the south and west and the River Colne to the east which can be used as known-head boundaries; and

(2) certain hydrogeological features to the north where the component of flow normal to the boundary is known (either zero or estimated from groundwater contours).

With the numerical model of the coastal aquifer (Fig. 7.3.2; the numbers on the elements are the local values of $\bar{K}_x$, $\bar{K}_y$ and the storativity), the region is extended so that a river can be used as a known-head boundary on one side and the northern edge taken as a no-flow boundary. The other sides are bounded by the sea.

Subdivision of the region into elements

The Lower Colne example uses four-node bilinear quadrilaterals. The coastal aquifer example uses eight-node isoparametric quadrilaterals. Quadrilaterals are a convenient shape of element to use for an aquifer model. The eight-node elements are more complicated but as explained in section 6.4 they can be used to approximate curved boundaries by parabolas.

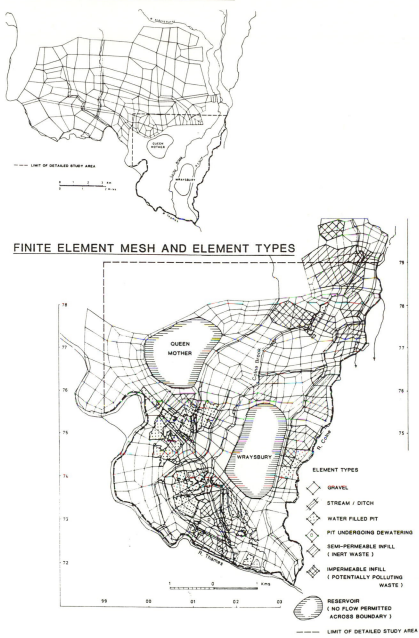

Figure 7.3.1: Finite element mesh for the Colne Gravels study (Connorton and van Beeston, 1983).

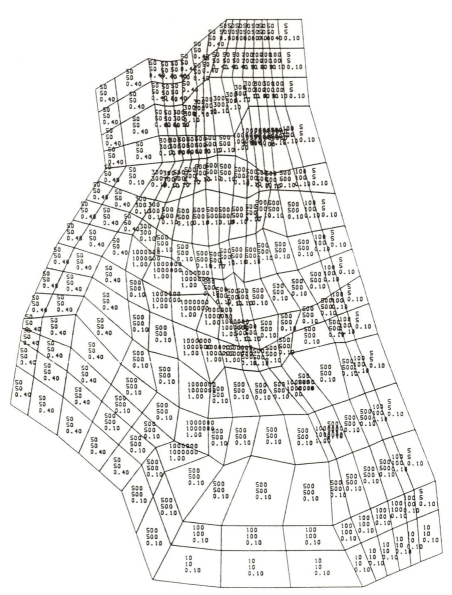

Figure 7.3.2: Finite element mesh for a coastal aquifer.

The position of nodes

1. The nodes should be positioned so that physical properties, such as storativity and hydraulic conductivity or transmissivity which vary areally, can be adequately represented.

2. Abstraction or recharge wells represented by point sinks or sources must be at finite element nodes. As with the finite difference equation obtained by the integration method (section 5.2) wells are conveniently represented by Dirac delta generalized functions in the term Q in eqn (7.2.1). A well at node s, positioned at (x_s, y_s), say, is represented by a term

$$\hat{Q}_s \delta(x - x_s, y - y_s) \tag{7.3.1}$$

where $\hat{Q}_s$ is the recharge rate from this well. This term only produces an effect in the equation which is formed by multiplying by $N_s(x, y)$, the basis function corresponding to node s (recall that a basis function is equal to unity at its own node and zero at all other nodes). Then integrating over the region R gives the result

$$\iint\limits_{R} \hat{Q}_s \delta(x - x_s, y - y_s) N_s(x, y) \, dx \, dy = \hat{Q}_s \tag{7.3.2}$$

because the Dirac delta generalized function has the property when integrated of filtering out the value of the integrand at (x_s, y_s).

3. The streams in the interior of the Colne model are 'leaky streams' modelled differently from the ephemeral streams discussed in section 5.5. In Jacoby (1982) they are enclosed in long thin elements as shown in Fig. 7.3.1. The leakage into the aquifer per unit area is given by

$$Q_s = \begin{cases} K_{sl}(h_s - h), & h \geq h_d \\ K_{sl}(h_s - h_d), & h < h_d \end{cases} \tag{7.3.3}$$

where h_s, h_d are the levels of the stream surface, and bed respectively, and the constant of proportionality K_{sl} has to be obtained by calibration.

4. It is convenient if calibration wells are also at nodes but this is not essential as the finite element solution can easily be recovered at any point.

5. Size of elements: the finer the mesh, i.e. the smaller the elements in general, the more accurate is the solution but also the more expensive to compute. A compromise is to solve with the coarsest

mesh which can reasonably define the problem and check the mass balance (section 7.5). If the error here is too great refine where the solution varies most.

6. Numbering of the nodes: the nodes must be numbered so as to minimize the bandwidth of the matrix which has to be inverted in the solution procedure. Gibbs *et al.* (1976) give an algorithm for reducing the bandwidth of a sparse matrix.

Water-filled gravel pits or lakes are represented by areas of very high hydraulic conductivity, $\bar{K}_x = \bar{K}_y = 10^6$ metres per day. Infilled pits are represented by areas of low permeability, $\bar{K}_x = \bar{K}_y = 5$ metres per day, or with some cases, 0.05 metres per day, which means they are effectively impermeable. The two large reservoirs in the Colne problem area are sealed off from the groundwater and have no-flow boundaries.

Since the gravel pits are being treated as part of the groundwater system, when water flows from a gravel pit into a channel it is leaving this system and must be represented as a sink in the model. When the water flows into the gravel pit from a channel it is a source. In the Lower Colne Gravels aquifer model the flow in the channel is supposed given by the Manning formula (section 1.2):

$$q = C^{-1/2}AR^{2/3}s^{1/2} \tag{7.3.4}$$

where A, R, and s are the cross-sectional area, the hydraulic radius, and the slope of the channel respectively and C is the constant of proportionality as in eqn (1.2.4). If the depth of the water in the channel is H and the channel can be taken to be of rectangular section with width b, then $A = bH$ and $R = A/(b + 2H)$. The value of the constant C has to be obtained from calibration but Jacoby (1982) reports that the effect of variation of C is only local. This is as expected with a parabolic problem (Chapter 1).

The term Q in eqn (7.2.1) must include the effective precipitation as well as the effects of abstraction or recharge wells, leaky streams, and water entering or leaving the aquifer via (non-leaky) channels as described above.

7.4 Aquifer models—initial conditions

For the time-dependent model of an aquifer, represented by the parabolic equation (7.2.1), initial conditions must be supplied (Chapter 1). From the results in Rushton and Wedderburn (1973) it appears that it is better to start from a dynamic balance (i.e. flow in equals flow out) steady state rather than from one where there is no flow at all which may anyway be unrealistic. Aquifer models like the Colne and the coastal aquifer described here are usually run with monthly time steps and an annual variation of data. (In dynamic vibration problems such as the numerical simulation of earthquake effects an engineer's rule of thumb says there should be at least 10 time steps per period.) This is repeated until a periodic solution is achieved and the model is then considered to be run in and ready to go on to model the particular developments to be studied. The coastal aquifer model required only 2 years' run-in from a dynamic balance steady state such as that shown in Fig. 7.4.1.

7.5 Mass balance

Equation (7.2.4) is the weak form (Chapter 6) of the steady-state groundwater flow equation (7.2.2). Suppose the specific discharge term $-\bar{K}(h - h_{\rm b})\nabla h$ (or $-\bar{T}\nabla h$ when the transmissivity approximation is used) is replaced by $\mathbf{q}$. Take the standard finite element Galerkin approach with w replaced by a typical basis function $N_j(x, y)$ and the double integrals replaced by integration formulae denoted by

$$\iint \mathbf{q} \cdot \nabla N_j(x, y) \, {\rm d}x \, {\rm d}y \approx \sum{}^{\rm I} \hat{\mathbf{q}} \cdot \nabla N_j(x, y) \tag{7.5.1}$$

where $\hat{\mathbf{q}}$ is now the vector of nodal values of the flux given by the finite element solution. Thus eqn (7.2.4) can be written as

$$-\sum{}^{\rm I} \hat{\mathbf{q}} \cdot \nabla N_j(x, y) + \int_{S_1+S_2} \hat{\mathbf{q}} \cdot \mathbf{n} N_j(x, y) \, {\rm d}S = Q_j \tag{7.5.2}$$

where Q_j represents the source or sink at node j. Sum the equations (7.5.2) over all the nodes j and use the result from section 6.1 that

$$\sum N_j(x, y) = 1, \quad \text{everywhere} \tag{7.5.3}$$

and hence

$$\sum \nabla N_j(x, y) = \nabla \sum N_j(x, y) = [0, 0]^{\rm T}. \tag{7.5.4}$$

Since in the approximation of the integral by a sum in (7.5.1), each value of a component of $\hat{\mathbf{q}}$ is multiplied by the sum of the components of $\nabla N_j(x, y)$ in its vicinity, the result is

$$\int_{S_1+S_2} \hat{\mathbf{q}} \cdot \mathbf{n} \, {\rm d}S = \sum_j Q_j \tag{7.5.5}$$

i.e. flow in equals flow out. Thus with the finite element solution

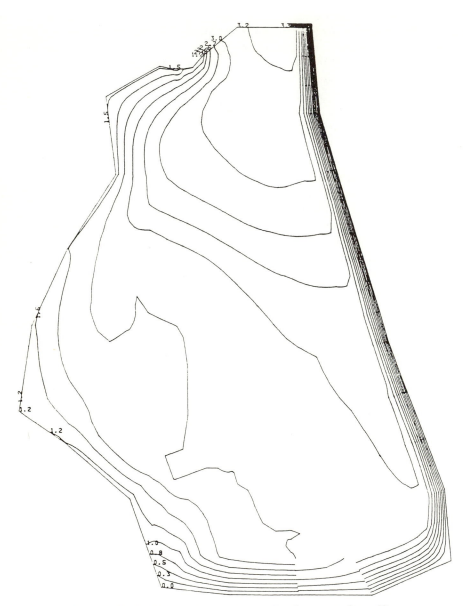

Figure 7.4.1: Steady-state solution for the coastal aquifer.

substituted into the known-head node equations the flow across the boundary here is obtained in such a way that there is exact mass balance (Lynch, 1984).

Another way of looking at this is to take eqn (7.5.2) written in the form

$$\sum k_{ij}\hat{h}_j + \int_{S_1+S_2} \hat{\mathbf{q}}\cdot\mathbf{n}N_j(x, y)\,\mathrm{d}S = Q_j \qquad (7.5.6)$$

where $\hat{h}_j$ is the finite element solution at node j. Then it can be seen that the fact that adding over all the nodes j results in zero contribution from the first term on the left of eqn (7.5.6) shows again that before any known-head nodal values have been overwritten the stiffness matrix is singular (adding the entries in each column gives zero). The best way to find the flow across the known-head boundary is to use the known-head boundary node equations as above. But calculations based on obtaining the gradient of the finite element solution at the known-head boundary can be useful in estimating how good the finite element solution is and whether the mesh needs to be refined. The following are ways of calculating the flow across the known-head boundary from the finite element solution for the three kinds of element most likely to be used for groundwater flow problems.

1. For linear triangles the gradients are most accurate at the midpoints of the sides (Levine, 1985). The flow across the side of a triangle on a known-head boundary can be taken as approximated by

$$\bar{K}\nabla\phi\cdot\mathbf{n} \text{ times the length of the side} \qquad (7.5.7)$$

in terms of the finite element solution for the potential ϕ, where $\nabla\phi = (\partial\phi/\partial x, \partial\phi/\partial y)^{\mathrm{T}}$, is the (constant) gradient from that element and $\mathbf{n} = (n_x, n_y)^{\mathrm{T}}$ is the outward normal.

Strang and Fix (1973) observed that for six-node triangles the midpoints of the sides seem likely to give superconvergent values for the derivatives along these sides. As mentioned in section 6.5 Levine (1985) proves this conjecture for six-node triangles which can be coupled in pairs to form parallelograms, i.e.

(a) the derivatives along the sides of the triangle are superconvergent at the midpoints and
(b) the average of the normal derivatives either side of a triangle edge are superconvergent at the midpoints.

Also the average of the gradients at three edge midpoints gives a superconvergent value at the centroid of the triangle. Levine also suggests that if the gradient is required at a vertex it should be calculated as an average of the midpoint gradients from the elements which share that vertex node. This last result can be used to calculate the flow across the side of a triangle on a known-head boundary using the trapezium rule.

2. For bilinear quadrilaterals the gradient is most accurate at the centroid of the element (Barlow, 1976). From section 6.6 for isoparametric quadrilaterals the components of the gradient of the finite element solution for the potential ϕ are given by

$$\frac{\partial \phi}{\partial x} = \frac{1}{|\mathbf{J}|} \sum_j \phi_j \left[\frac{\partial N_j}{\partial \xi} \frac{\partial y}{\partial \eta} - \frac{\partial N_j}{\partial \eta} \frac{\partial y}{\partial \xi} \right] \tag{7.5.8}$$

and

$$\frac{\partial \phi}{\partial y} = \frac{1}{|\mathbf{J}|} \sum_j \phi_j \left[-\frac{\partial N_j}{\partial \xi} \frac{\partial x}{\partial \eta} + \frac{\partial N_j}{\partial \eta} \frac{\partial x}{\partial \xi} \right]. \tag{7.5.9}$$

At the point $\xi = \eta = 0$ the gradient is obtained with the same order of accuracy as the finite element solution itself, i.e. with error $O(h^2)$. If the elements are arranged so that their sides are parallel to the boundary as in Fig. 7.5.1, the values of the gradients sampled at centroids A, B in Fig. 7.5.2 can be extrapolated to give a value at P, the midpoint on the side on the known-head boundary. The flow across the boundary is then estimated as in (7.5.7).

It is slightly awkward that the numerical integration for the bilinear quadrilateral element should in general use the 2×2 Gauss point formula, not the one-point centroid sampling. For parallelograms the

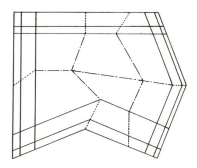

———————— groups of three lines adjacent to the region boundary, which must be parallel, but only piecewise straight.

– – – – – – – – – lines which must be straight.

– · — · — · — · lines having no special restrictions.

Figure 7.5.1: Finite element mesh showing edge details (Matthews and Wood, 1987).

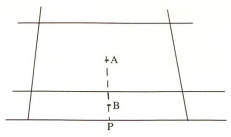

Figure 7.5.2: Extrapolation to boundary using centroids.

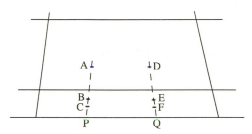

Figure 7.5.3: Extrapolation to boundary using 2 × 2 Gauss points.

values of the gradient at the centroid can be obtained by taking the average of the values at the 2 × 2 Gauss points. Alternatively this requires a small additional piece of program.

3. For the eight-node quadrilaterals as used in the coastal aquifer model (Fig. 7.3.2) the optimum points at which to sample the gradients are the 2 × 2 Gauss points (Barlow, 1976). At these points the gradients are obtained with the same order of accuracy as the finite element solution, i.e. with error $O(h^3)$. Hence with the element sides aligned as in Fig. 7.5.1 taking parabolas through A, B, C and D, E, F in Fig. 7.5.3 gives values of the gradient at points P, Q on the boundary. These are then at the Gauss points there ready to integrate to find the flow across the known-head boundary. Matthews and Wood (1987) take a numerical simulation of a phreatic aquifer (two dimensions in plan) with or without an extraction well using the eight-node elements as in the coastal aquifer model. The mass-balance error which is used as an indicator of a sufficiently refined finite element mesh is taken as

$$e\% = \frac{100|\text{flow in} - \text{flow out}|}{\text{throughput}} \qquad (7.5.10)$$

where

$$\text{throughput} = \tfrac{1}{2}(|\text{flow in}| + |\text{flow out}|). \qquad (7.5.11)$$

Thus for the problem considered the mass-balance error is given by

$$e\% = \frac{100|Q_w + T\int \nabla h \cdot \mathbf{n}\, dS|}{\frac{1}{2}(|Q_w| + T\int |\nabla h \cdot \mathbf{n}|\, dS)} \qquad (7.5.12)$$

where Q_w is the rate of abstraction from the well in the second example. Figure 7.5.4 (from Matthews and Wood, 1987) shows results from a kind of patch test where known heads from a known solution are applied round the boundary of a patch of elements in which the mesh is successively refined. A comparison is made between the results given by the mass-balance error for the flow across the boundary given by extrapolation and by direct sampling at the boundary. In Fig. 7.5.4 the percentage error in the mass balance, given in (7.5.10), is plotted against the ratio of the length of the region to the maximum element length on a log–log scale. With the regular mesh (rectangular element with sides parallel to the global x, y axes), the slopes are -2, as expected with the direct sampling, and -4 with extrapolation, which is better than expected. With an irregular mesh the two methods each give a slope of about -3.5 for the coarser meshes but the direct sampling is worse as the mesh is refined. Figure 7.5.5 (from Matthews and Wood, 1987) shows results with an irregular mesh for the case of the logarithmic function representing a well (section 1.8). The direct sampling is again worse as the mesh is refined. Matthews and Wood show that the extrapolation method here always requires less CPU time than the direct method for a given accuracy.

Numerical modelling of a barrier

A coastal aquifer may have a barrier inserted to the full depth of the aquifer in order to hold back the salt water. If the barrier is of significant breadth it may be represented by long thin elements. Otherwise it can be modelled by giving the nodes different numbers on either side of the barrier. This has been successfully used with the coastal aquifer model (Fig. 7.3.2). It is topologically equivalent to the reservoirs in the Lower Colne Gravels model (Figure 7.3.1).

7.6 Modelling a phreatic surface

For the three-dimensional case it is possible to solve for the piezometric head ϕ the equation

$$S_0 \frac{\partial \phi}{\partial t} = \frac{\partial}{\partial x}\left(K_x \frac{\partial \phi}{\partial x}\right) + \frac{\partial}{\partial y}\left(K_y \frac{\partial \phi}{\partial y}\right) + \frac{\partial}{\partial z}\left(K_z \frac{\partial \phi}{\partial z}\right) \qquad (7.6.1)$$

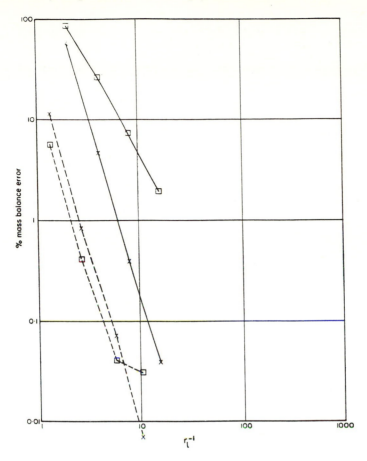

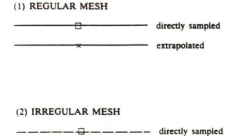

(1) REGULAR MESH

directly sampled

extrapolated

(2) IRREGULAR MESH

directly sampled

extrapolated

Figure 7.5.4: Percentage error in mass balance without abstraction well (Matthews and Wood, 1987).

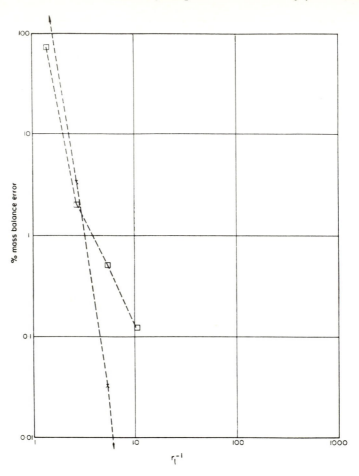

IRREGULAR MESH

Figure 7.5.5: Percentage error in mass balance with abstraction well (Matthews and Wood, 1987).

obtained from (A1.1) by substituting for q_x, q_y, q_z using Darcy's law. When the region is supposed saturated up to the phreatic surface and to have no water above that, the difficulty is that the position of this surface is initially unknown. As explained in Appendix 1 there are two boundary conditions to be satisfied on the phreatic surface: eqn (A1.7) and $\phi = z$. Taking a two-dimensions-in-elevation example with the assumptions of steady state and no accretion, the phreatic surface is a stream line, i.e. there is no flow in the direction of the outward normal here. This is simple to implement with finite elements. An initial position of the surface is estimated. Whether triangles or quadrilaterals are used the nodes are placed on verticals. The equations are solved with the no-flow condition on the surface. The heights z_j of the surface nodes j are adjusted until $\phi_j = z_j$ here to some tolerance. If the position of the surface oscillates it is a good idea to average to smooth convergence, i.e. if z_j^n is the height of node j at the nth iteration, and the resultant nodal value is $\phi_j^n = z_j^n + a$, put $z_j^{n+1} = z_j^n + a/2$.

7.7 The numerical solution of Richards' equation

Richards' equation combines flow in the saturated and unsaturated regions in the three-dimensional region up to the surface of the ground. Taking the coordinates as x, y horizontally and z vertically upwards, the general three-dimensional equation introduced in section 1.7 is now used in the form

$$C(\psi)\frac{\partial\phi}{\partial t} = \frac{\partial}{\partial x}\left[K_x(\psi)\frac{\partial\phi}{\partial x}\right] + \frac{\partial}{\partial y}\left[K_y(\psi)\frac{\partial\phi}{\partial y}\right] + \frac{\partial}{\partial z}\left[K_z(\psi)\frac{\partial\phi}{\partial z}\right] + \tilde{Q} \quad (7.7.1)$$

where $C(\psi) = \mathrm{d}\theta/\mathrm{d}\psi$ is the specific water storativity or the specific water capacity of the saturated/unsaturated medium, θ being the soil moisture content by volume (i.e. θ is dimensionless) and the capillary potential $\psi = \phi - z$. Typical relationships of θ as a proportion of the saturated value θ_s and the hydraulic conductivity K as a proportion of the saturated value K_s to the value of ψ are shown in Fig. 7.7.1 (data taken from Clapp and Hornberger, 1978). The term $\tilde{Q}$ represents source or sink terms; its dimensions are T^{-1}, i.e. the rate at which water enters or leaves ($\mathrm{L}^3\mathrm{T}^{-1}$) per unit volume ($\mathrm{L}^3$).

The discussion which follows here is mainly based on the author's experience with the IHDM (Institute of Hydrology Distributed Model, described in Beven *et al.* (1987)). Other examples of subsurface flow solved by finite element methods are given in Chapter 9 on water-quality modelling. The IHDM is a rainfall–runoff numerical model of a catchment which uses two-dimensional vertical sections of the catchment hillslopes with a variable slope width, B. These hillslopes can then be combined with a network of channels. With no source or sink term the groundwater flow is represented by the equation

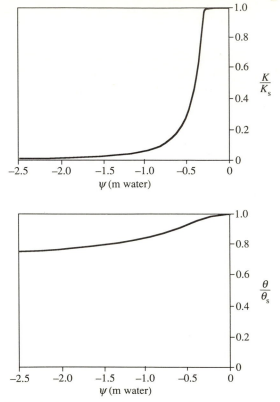

Figure 7.7.1: K/K_s and θ/θ_s against ψ (data from Clapp and Hornberger (1978)).

for $\phi(x, z, t)$:

$$BC(\psi)\frac{\partial\phi}{\partial t} = \frac{\partial}{\partial x}\left[BK_x(\psi)\frac{\partial\phi}{\partial x}\right] + \frac{\partial}{\partial z}\left[BK_z(\psi)\frac{\partial\phi}{\partial z}\right] \qquad (7.7.2)$$

assuming no variations in the y direction.

This groundwater flow problem is solved by finite elements as described in Chapter 6 using four-node bilinear elements as in Fig. 7.7.2 (from Wood and Calver, 1992). Figure 7.7.2 shows soil of a constant depth and parallelogram elements but a variable depth and more general elements can be used. The boundary conditions are known head or known flow as follows:

1. Known head at the nodes adjacent to the channel at the foot of the slope.

2. Zero flow at the drainage divide at the top of the slope. Frequently

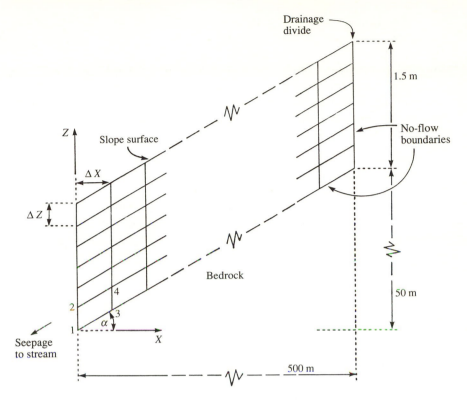

Figure 7.7.2: Finite element mesh for hillslope (Wood and Calver, 1992).

the bottom boundary of the region is not completely impervious and any leakage can be included as a 'known-flow' boundary condition.

3. On the surface of the soil there is the result of the effective precipitation (precipitation minus evaporation and interception by plants):

(a) If the soil surface is unsaturated the effective precipitation enters the soil at the given rate (known flow). The net precipitation over the element surface equals the precipitation rate times the total surface area of the element. In the calculation of element contributions in the finite element program this is assigned to the nodes according to the slope area associated with each node.

(b) If the soil surface is saturated then the effective precipitation is added to the kinematic wave equation representing the surface flow which is solved by finite differences (section 4.2). The upper boundary for the overland flow is usually a zero-flow boundary as in the example in Chapter 2. The solution of the kinematic wave equation then gives the overland flow on the surface of the slope.

When the surface is saturated the surface boundary condition for the groundwater flow is changed to a known-head boundary condition with the potential at atmospheric, i.e. $\phi = z$. (The small depth of overland flow is neglected.) There may be a flow of water out from the saturated surface to join the overland flow. This is calculated by the Lynch method described in section 7.2.

4. On the potential seepage face at the bottom of the slope seepage occurs below where the saturated zone intersects this face. The point where this occurs has to be found by iteration. Above the seepage face the soil is unsaturated and this is treated as a zero-flow boundary in the solution of the equations. On the seepage face the soil is saturated; hence the pressure is atmospheric and $\phi = z$, and also the flow is outwards. It is best to start the iteration with the seepage nodes too high and test from the bottom upwards. The seepage face nodes are tested until one is found with an influx greater than zero (calculated as in section 7.2) and then this and those higher up are changed to potential seepage face nodes. It is also possible to test the seepage face nodes independently to allow for a vertical variation in the hydraulic conductivity.

Wells in the interior of a three-dimensional region whose dimensions are small compared with those of the region can be represented by 'line sources or sinks' along vertical edges of elements. This is an extension of the device of using the Dirac delta generalized function to represent wells in aquifer problems. The IHDM, which is two dimensions in elevation with a variable breadth, assumes no variation in the y direction and thus is not suitable for the inclusion of wells.

A rainfall–runoff model such as the IHDM is designed to study the effects of physical controls of runoff. Hence the time scale may be hours rather than years as with the response of an aquifer to annual cycles. Suitable initial conditions for a hillslope model are more difficult to determine than for an aquifer. Initial conditions are stated for the particular examples which follows and are discussed further in section 7.9.

The result of the finite element discretization of Richards' equation (7.7.2) is a system of ordinary differential equations of the form

$$\mathbf{M}(\phi)\,\frac{\mathrm{d}\phi}{\mathrm{d}t} + \mathbf{K}(\phi)\phi = \mathbf{F}(t) \tag{7.7.3}$$

where $\mathbf{M}(\phi)$ is the mass matrix

$$m_{ij} = \iint BC(\psi)N_i N_j \, \mathrm{d}x \, \mathrm{d}z \qquad (7.7.4)$$

and $\mathbf{K}(\phi)$ is the stiffness matrix

$$k_{ij} = \iint B\left[K_x(\psi) \frac{\partial N_i}{\partial x} \frac{\partial N_j}{\partial x} + K_z(\psi) \frac{\partial N_i}{\partial z} \frac{\partial N_j}{\partial z} \right] \mathrm{d}x \, \mathrm{d}z. \qquad (7.7.5)$$

Because of the presence of the terms $C(\psi)$, $K_x(\psi)$, $K_z(\psi)$ the mass and stiffness matrix entries depend on the solution $\phi = \psi + z$ (Fig. 7.7.1). Hence the system of equations (7.7.3) is non-linear.

The time derivative in eqn (7.7.3) is approximated using the backward finite difference (Chapter 3) which is unconditionally stable and gives a smooth result:

$$\left[\mathbf{K}(\bar{\phi}) + \frac{1}{\Delta t} \mathbf{M}(\bar{\phi}) \right] \phi_{n+1} = \bar{\mathbf{F}} + \frac{1}{\Delta t} \mathbf{M}(\bar{\phi}) \phi_n \qquad (7.7.6)$$

where $\bar{\mathbf{F}}$ is an average value of $F(t)$ and $\bar{\phi}$ an average value of ϕ over the time interval Δt. ϕ_n is the approximate value of the nodal values of ϕ at time t_n. Equation (7.7.6) can be written in the form

$$\mathbf{A}\phi_{n+1} = \mathbf{H} \qquad (7.7.7)$$

where the entries in the matrix $\mathbf{A}$ and the vector $\mathbf{H}$ depend on the value of ϕ_{n+1}. Allowance for an adjustable time step size is made by putting $t_{n+1} = t_n + \Delta t_n$. Iteration on all time steps except the first begins by the extrapolation:

$$\bar{\phi}_{n+1} = \phi_n + 0.5 \frac{\Delta t_n}{\Delta t_{n+1}} (\phi_n - \phi_{n-1}). \qquad (7.7.8)$$

The iteration is then accelerated by an adaptation of the Cooley algorithm (Cooley, 1983). The equations are solved for the change $\Delta\phi_{n+1}^k$ in the value of the potential from one iteration to the next, where

$$\Delta\phi_{n+1}^k = \phi_{n+1}^k - \phi_{n+1}^{k-1}. \qquad (7.7.9)$$

The Cooley algorithm assumes that these changes will oscillate in sign from one iteration to the next so that the convergence is accelerated by damping this oscillation. This is equivalent to underrelaxation. If Δ_k is the maximum change in nodal potentials at the kth iteration, the method uses

(1)
$$s = 1, \quad \text{for } k = 0$$
$$s = \Delta_{k+1}/(\omega_k \Delta_k), \quad \text{for } k > 0 \qquad (7.7.10)$$

where ω_k is the relaxation parameter from the previous iteration;

(2) $\hat{\omega} = (3 + s)/(3 + |s|)$,   when $s > -1$ (7.7.11)

otherwise $\hat{\omega} = 0.5/|s|$;

(3) if $\hat{\omega}|\Delta_{k+1}|$ is greater than e_{max}, the specified maximum change in the potential to be allowed in an iteration, then

$$\omega_{k+1} = e_{max}/|\Delta_{k+1}|$$ (7.7.12)

otherwise $\omega_{k+1} = \hat{\omega}$.

The iteration stops when $|\Delta_{k+1}|$ is less than the prescribed tolerance.

This adaptation of the Cooley algorithm works very efficiently with the IHDM model. More iterations are needed at the start of the run and when the rainfall starts but for most time steps only a few iterations are necessary. With an example involving 25 hours without rain, 5 hours with steady 4 mm h^{-1}, and another 25 hours without rain, and a tolerance of 0.001 m of water, an average of between two and three iterations per 0.5 h time step is required and an average of one to two iterations for a time step of 0.1 h. The total number of iterations differs little between the different grids described in the next section.

7.8 Choice of hillslope finite element grid

Experiments with a range of finite element grids for the numerical solution of Richards' equation for subsurface flow in a hillslope as shown in Fig. 7.7.2 are described in this section. This hillslope consists of 1.5 m depth permeable soil overlying an impermeable bedrock. There is 500 m horizontal distance between the divide and the channel at the foot and constant 0.1 slope with straight contours. The details of the other parameters used are to be found in Calver and Wood (1989). In such a long thin region it is natural to consider using long thin elements but what ratio, length:height, is acceptable? In these experiments the results are compared with those given by a baseline grid of equal parallelogram elements with $\Delta x = \Delta z = 0.5$m, where Δx, Δz are the horizontal and vertical dimensions respectively. This uses 4004 nodes and is expensive to run. The experiments aim to discover how much coarser the mesh can be and still give useful results.

Table 7.8.1 summarizes the results of the experiments including the mass-balance error as calculated by the Lynch method. Figure 7.8.1 shows the discharge per metre hillslope width at the foot of the slope for the different grids. All the runs take the first 25 h as run-in, then 5 h rain (at 4 mm h^{-1}) followed by a further 25 h with no rain. During the first 25 h with no rain there should be a smooth run-in to a very

Table 7.8.1 Mass balance error calculated by the Lynch method

Run	Horizontal dimension of elements Δx (m)	$\Delta x/\Delta z$	Total no. of nodes	CPU time (s)	Total no. of iterations	Mass balance (m^3)
1a	0.5	1	4004	1407	253	0.005 30
2a	100	200	24	9	230	−0.000 53
3a	20	200	416	258	252	0.011 07
4a	50	100	44	15	229	0.001 61
5a	25	100	147	64	249	0.010 33
6a	10	100	816	523	253	0.009 45
7a	10	20	204	77	252	0.006 10
8a	5	20	707	314	251	0.008 69
9a	2	20	4016	2327	253	0.009 69
10a	5	10	404	145	254	0.004 67
11a	2.5	10	1407	629	252	0.009 20
12a	100, 10*	200, 20	60	20	226	0.003 54
13a	variable†		342	205	226	0.004 42

* Lower fifth of hillsope has the finer mesh.
† Depths (m) of node rows from surface: 0.00 0.05 0.10 0.15 0.20 0.30 0.40 0.50 0.60 0.70 0.80 0.90 1.00 1.10 1.20 1.30 1.40 1.50.
Distances (m) of node columns from divide: 0 100 200 300 400 410 420 430 440 450 460 470 480 490 492 494 496 498 500.

slowly varying discharge at the foot of the slope. Figure 7.8.1 shows the spurious initial peak in the flow due to the elongated elements in the coarse grid.

Two measures of the deviation from the baseline run are:

1. The ϕ deviation

$$\left[\frac{1}{N_n} \sum_{j=1}^{N_n} (\phi_j^b - \phi_j)^2 \right]^{1/2} \tag{7.8.1}$$

calculated at the end of the run ($t = 55$ h), where ϕ_j^b are the baseline run values of the hydraulic potential and N_n is the number of nodes (j) common to all the runs.

2. The Q deviation

$$\left[\frac{1}{N_t} \sum_{k=1}^{N_t} (Q_k^b - Q_k)^2 \right]^{1/2} \tag{7.8.2}$$

where Q_k^b are the baseline values of the discharge at the foot of the slope and N_t the number of 0.5 h time steps.

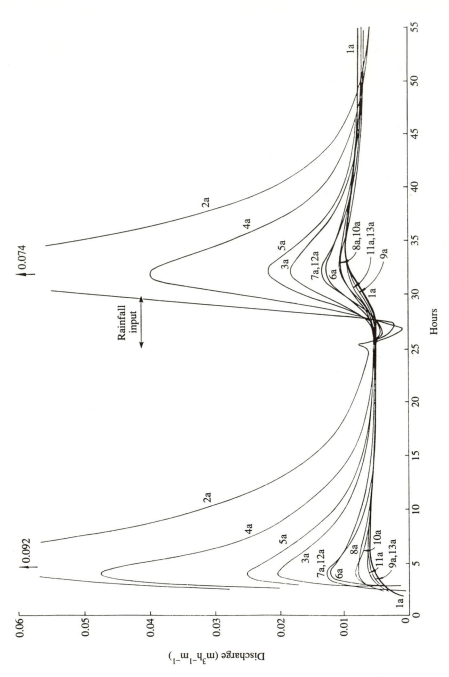

Figure 7.8.1: Discharge at foot of hillslope (Wood and Calver, 1992).

Figures 7.8.2 and 7.8.3 show the ϕ deviation and the Q deviation for the different grids defined in Table 7.8.1. The non-regular grids with smaller elements near the foot of the slope have comparatively high ϕ deviations for their total node number but good values for the Q deviation. This is because of the smaller elements near the seepage face where the discharge is calculated. The results with $\Delta t = 0.1$ h are very similar.

Results for the mass-balance error calculated by the Lynch method in Table 7.8.1 show that this is no use for indicating which grid is best. Runs such as 2a give an apparently good result for mass balance but large ϕ and Q deviations and very bad overshoot in the hydrograph in Fig. 7.8.1. The Lynch method used with the equations resulting from the time-stepping scheme (7.7.6) gives the flow out across the known-head boundary in the vicinity of node j as

$$-\frac{1}{\Delta t} \sum_i m_{ij}(\phi_{i,n+1} - \phi_{i,n}) + \sum_i k_{ij}\phi_{i,n+1} + f_j. \tag{7.8.3}$$

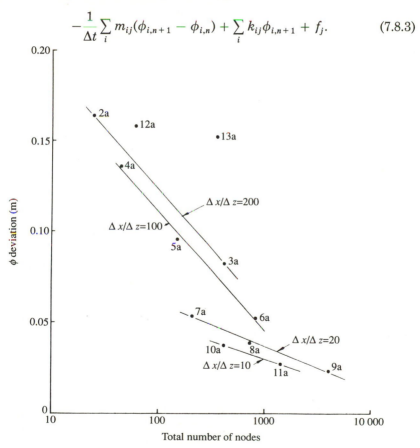

Figure 7.8.2: ϕ deviation (Calver and Wood, 1989).

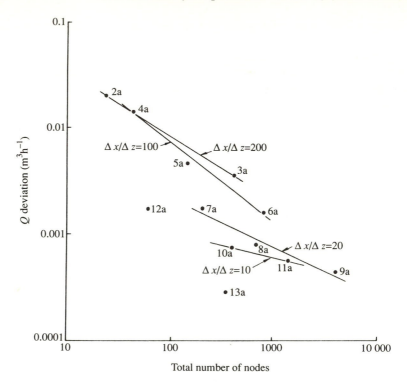

Figure 7.8.3: Q deviation (Calver and Wood, 1989).

As shown in section 3.2 the error due to this approximation of the time derivative for reasonably small values of the time step Δt is proportional to the second derivative of the ϕ solution. Thus the erratic values of the mass-balance error as calculated by the Lynch method appear to be explained as follows.

The fixed head node at the base of the seepage face remains at the same fixed level throughout the run. Hence the change in the discharge at this fixed head node is due to the change in the values of ϕ in the neighbourhood of this node. The spurious peak in the discharge in run 2a corresponds to the spurious peak in the ϕ values which is due to the very coarse grid. The error in the run 2a is proportional to the second time derivative of the ϕ solution as Δt tends to zero but the finite element grid remains fixed. Thus the cumulative mass-balance error as given by the Lynch formula is the result of adding terms with changing sign of $\partial^2 \phi / \partial t^2$ and this makes it lower than the results using other grids where the second time derivative behaves correctly.

Figure 7.8.4 shows two measures of the success of the discharge prediction plotted against CPU time for the 0.5 h time step runs. These

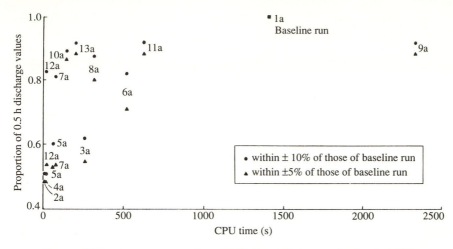

Figure 7.8.4: Accuracy against CPU time (Calver and Wood, 1989).

discharge errors are calculated after 25 h of run, i.e. from the beginning of the rainstorm, to exclude the run-in effects. The two measures plotted are the percentage of the half-hourly values within $\pm 10\%$ and $\pm 5\%$ of the values given by the baseline run. In practical use the numerical model is likely to be calibrated against discharge; hence this is a good method for judging cost-effectiveness. Figure 7.8.4 suggests that for cost-effectiveness grids 7a, 8a, 10a, 12a and 13a are best at the $\pm 10\%$ level. These have grids with $\Delta x/\Delta z$ not greater than 20 or non-regular mesh as specified in Table 7.8.1 and total node number less than about 750 (which averages about one node per square metre of the vertical section of hillslope). The best runs at the $\pm 5\%$ level are runs 8a, 10a, and 13a. These have $\Delta x/\Delta z$ values of 10, 20, and variable but more nodes than 7a and 12a.

The conclusions from these results are:

1. If a hydrograph such as Fig. 7.8.1 is plotted the spurious kick in the discharge during the run-in indicates a bad choice of grid. Elements too large is one source of spurious oscillation, 'noise'.

2. Better values of the discharge are obtained by mesh refinement near the seepage face. Hence refine here and gradually increase the size of the elements up the slope as much as expense warrants (i.e. limit of CPU time is reached) until the solution is reasonably smooth.

3. With this procedure neither the number of iterations required nor the cumulative mass balance as calculated by the Lynch method help in choosing the best grid.

There is confusing advice in the literature as to whether a lumped (i.e. diagonal) or distributed (as in (6.2.2)) mass matrix should be used for the solution of Richards' equation. Wood and Calver (1990) discuss this and conclude from experiments with the IHDM that it makes a significant difference to the discharge from the slope. This difference may be considerably in excess of field measurement error and the use of a distributed mass matrix is strongly recommended.

7.9 Initial conditions for a hillslope model

Time-dependent parabolic equations such as (7.2.1) which models an aquifer or (7.7.2) which models the flow in a hillslope must have initial values of the hydraulic potential with which to start. Suitable initial conditions for a hillslope are more difficult to determine than for an aquifer. As stated in section 7.4 for an aquifer it is suitable to start from a steady state, either zero flow or a 'flow in equals flow out' dynamic balance. With a hillslope zero flow means that the total hydraulic potential ϕ which equals $\phi + z$ is constant everywhere and this is not realistic in a humid temperate region. It is also not possible to establish a precisely steady state in which inflow at the surface equals outflow at the bottom of the slope except when there is no water left. This is because of the continual changes taking place in the saturated/unsaturated regions. The best that can be achieved is with a discharge that shows a smooth decline to a slowly changing value. The numerical model can then be regarded as 'run-in' and the rainfall event introduced. As shown in the graphs of the discharge Q from the foot of the slope in Fig. 7.8.1 with a reasonable mesh such as 1, 11, or 8 any numerical noise will die away in a few time steps. There should then be a smooth decline to a slowly changing value and there should not be any violent oscillation when the rainfall starts. The results with coarser grids do not improve with longer run-in times. With a suitable grid large time steps can be used in the run-in period.

Wood and Calver (1992) experiment with two kinds of initial conditions using mesh 1 or mesh 8. The following is a summary of the consequences:

1. Taking the initial moisture potential $\psi_{in} = $ constant, i.e. $\phi = z + $ constant (u values in Figs 7.9.1 and 7.9.2, from Wood and Calver (1992)) gives a general vertical hydraulic gradient; a saturated layer develops all the way up the slope; the discharge Q is progressively higher with successive rainfall events (Fig. 7.9.1); and the time before the first rainfall is less critical (Fig. 7.9.2).

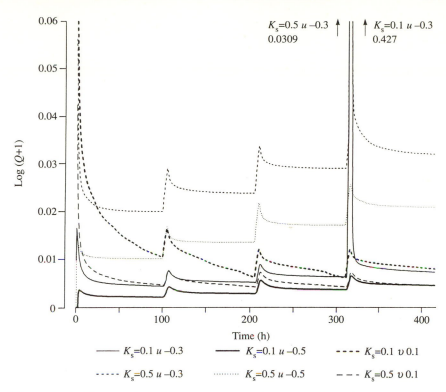

Figure 7.9.1: Effect of initial pressure potential on hillslope discharge (Wood and Calver, 1992).

2. Taking the initial moisture potential ψ_{in} as a fraction of the height above the bottom of the slope, e.g. $\psi_{in} = 0.1z$ (v values in Figs 7.9.1 and 7.9.2), a saturated wedge develops at the foot of the slope; the discharge Q is approximately the same or decreasing with repeated rainfall events (Fig. 7.9.1); and the time before the first rainfall event is more critical (Fig. 7.9.2).

The saturated layer which soon develops with the first case, $\psi_{in} =$ constant, if left long enough with no rainfall develops into a wedge as shown in Fig. 7.9.3 (from Wood and Calver, 1992). Table 7.9.1 (Wood and Calver, 1992) shows that after a period the ψ_{in} values make very little difference to the discharge Q from the foot of the slope. This period is shorter with the higher K_s values; the effect of the precise ψ_{in} values persists longer with lower K_s. These results give some guidance as to the choice of the initial conditions in order to produce a sequence of events more like (1) or more like (2).

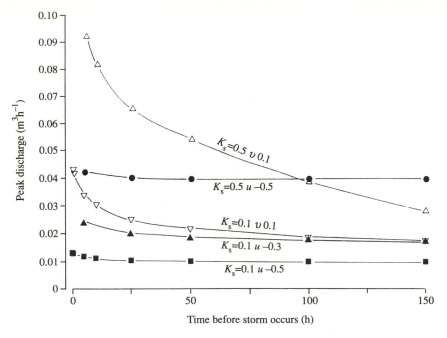

Figure 7.9.2: Peak discharge against time before storm occurs (Wood and Calver, 1992).

It is also possible to make an estimate of the appropriate initial ϕ distribution in the hillslope from a knowledge of (a) the hillslope discharge Q which is assumed to be reasonably steady and not affected by recent rainfall and (b) the saturated hydraulic conductivity K_s.

The numerical examples which follow are from Wood and Calver (1992) using the IHDM applied to the hillslope with dimensions as shown in Fig. 7.7.2 and the assumption of an isotropic medium. The soil is assumed to be of a uniform loam texture and its moisture and conductivity characteristics are detailed in Wood and Calver (1990). In these examples the level of the water in the channel is taken to be at or below the level of the lowest node on the seepage face.

Darcy's law

This approximation assumes there is a quantity of water Q (equal to the hillslope discharge) flowing down the slope, $\tan \alpha$, in a saturated layer of depth d and hence

$$Q = dK_s \tan \alpha. \tag{7.9.1}$$

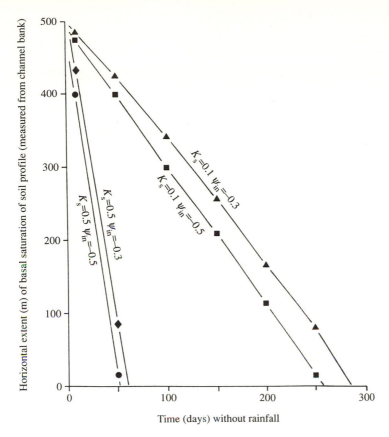

Figure 7.9.3: Horizontal extent of basal saturation against time without rainfall (Wood and Calver, 1992).

Table 7.9.1 Slope–base discharges during drainage period for different initial conditions. Q in m^3 per hour for unit width; subscript denotes length of drainage period in days

ψ_{in} (m)	K_s (m h^{-1})	Q_{100}	Q_{300}	Q^*_{1000}
-0.5	0.1	0.004 67	0.002 25	0.000 48
-0.3	0.1	0.009 35	0.002 70	0.000 48
-0.5	0.5	0.004 23	0.001 64	0.000 54
-0.3	0.5	0.004 30	0.001 65	0.000 54

* To five decimal places.

With $\tan \alpha = 0.1$, $K_s = 0.1$ metres per hour, and a unit width slope discharge of 0.004 55 cubic metres per hour (7.9.1) gives the depth of saturated layer as 0.455 metres. But in the numerical model the saturated depth over most of the hillslope is 0.15 metres. So using Darcy's law in this way seems simplistic.

Reverse Lynch

The idea here is to use the known value of the discharge Q to deduce the values of ϕ in the bottom slope element by using the Lynch formula (7.8.3) in reverse. The flow needs to be reasonably steady so that the storage term from the mass matrix in (7.8.3) is small compared with the Darcy term (see Wood and Calver (1990) for details). The formula for the discharge Q when only the bottom node is saturated is then

$$Q \approx k_{12}\phi_2 + k_{13}\phi_3 + k_{14}\phi_4 \tag{7.9.2}$$

where k_{12}, k_{13}, k_{14} are the entries in the stiffness matrix. For the parallelogram elements with node numbers as shown in Fig. 7.7.2 these entries are given by

$$\left.\begin{array}{l} k_{12} = K\left[\dfrac{\Delta z}{6\Delta x} - \dfrac{\Delta x}{3\Delta z}(1 + \tan^2 \alpha)\right] \\[3mm] k_{13} = K\left[-\dfrac{\Delta z}{3\Delta x} + \dfrac{\Delta x}{6\Delta z}(1 + \tan^2 \alpha)\right] \\[3mm] k_{14} = K\left[-\dfrac{\Delta z}{6\Delta x} - \dfrac{\Delta x}{6\Delta z}(1 + \tan^2 \alpha) + \dfrac{\tan \alpha}{2}\right] \end{array}\right\} \tag{7.9.3}$$

where K is the local average of the hydraulic conductivity and α is the slope angle. The flow is assumed to be predominantly horizontal so that $\phi_1 \approx \phi_4$ and it is essential that k_{13} and k_{14} are of the same sign, i.e. both negative, otherwise this method produces some very confusing results. To make $k_{13} < 0$ requires

$$-\frac{\Delta z}{3\Delta x} + \frac{\Delta x}{6\Delta z}(1 + \tan^2 \alpha) < 0 \tag{7.9.4}$$

i.e. $\Delta x < 1.4\Delta z$, approximately, with $\tan \alpha = 0.1$. Hence it is necessary to use $\Delta x = \Delta z$ in the bottom element. The entries in the stiffness matrix are then

$$k_{12} = -0.17K_s, \quad k_{13} = -0.165K_s, \quad k_{14} = -0.285K_s \tag{7.9.5}$$

assuming also that K_s is the average value of the hydraulic conductivity in the bottom element. Then assuming $\phi_3 = \phi_4$ and that ϕ_2 can be neglected, eqn (7.9.2) gives

$$\phi_3 = \phi_4 = \frac{|Q|}{0.45K_s} \qquad (7.9.6)$$

(since the discharge Q is negative in these examples because it is in the negative x direction it is simpler to use its numerical value, $|Q|$, here). Some numerical results from using this are shown in Table 7.9.2 (Wood and Calver, 1992). A disadvantage of this method is that apart from the bottom element the rest of the values of ϕ have to be put in by guesswork.

Dupuit–Forchheimer parabola approximation

The D–F (Dupuit–Forchheimer) assumptions are frequently used in unconfined groundwater flow (see Bear, 1979, Chapter 4). These suggest that the form of the saturated/unsaturated boundary near the base of a hillslope can be approximated by a parabola. This parabola intersects the potential seepage face at the level of saturation which depends on the channel water level. When the channel water level is at or below the base of the seepage face, this parabola is tangential to the seepage face at its base (node 1 in Fig. 7.7.2). The equation of the parabola is then

$$d^2 = 2\frac{|Q|}{K_s}x \qquad (7.9.7)$$

The D–F parabola meets the bottom of the soil where $d = x \tan \alpha$, i.e. at

$$x = 2\frac{|Q|}{K_s \tan^2 \alpha}. \qquad (7.9.8)$$

When using this D–F parabola approximation to estimate the initial values of ϕ in the slope it is essential to make sure the parabola does not intersect the soil surface (otherwise the water comes out of the top of the soil at the first time step). If l is the depth of the soil, the equation for the top of the soil in the examples considered in this section is

$$z = l + x \tan \alpha. \qquad (7.9.9)$$

This does not intersect the parabola if

$$|Q| < 2lK_s \tan \alpha. \qquad (7.9.10)$$

The D–F parabola gives the position of the point P, height d_P, where $\psi = 0$ on the line $x = \Delta x$ through nodes 3 and 4 in Fig. 7.7.2:

$$d_\mathrm{P} = \sqrt{(2|Q|\Delta x/K_\mathrm{s})}. \tag{7.9.11}$$

Assuming as before that $\phi_3 = \phi_4$, linear interpolation gives

$$\phi_3 = \phi_4 = d_\mathrm{P}. \tag{7.9.12}$$

This gives an estimate of the $\psi = 0$ position and reverse Lynch can now be used including the $k_{12}\phi_2$ term which was previously neglected:

$$k_{12}\phi_2 = Q - (k_{13} + k_{14})d_\mathrm{P}. \tag{7.9.13}$$

Taking this 'water table' parabola as approximating the boundary of the $\psi = 0$ region gives a result which is too high. It seems more reasonable to take it as approximating the top of the region corresponding to $\psi < 0$, $K = K_\mathrm{s}$, and low $\mathrm{d}\theta/\mathrm{d}\psi$ which includes the capillary fringe where there is only very limited air in the soil. With the relationship between K and K_s used in these examples (Fig. 7.7.1, details in Wood and Calver (1990)) an average value corresponds approximately to $\psi = -0.1$ m. This agrees with the values given by Bear (1979), quoting Silin-Bekchurin (1958), for the height of capillary fringe ranging from 2–5 cm in coarse sand to 2–4 m and more in clay. The result here is that the saturated layer meets $x = \Delta x$ at $d_a = d_\mathrm{P} - 0.1$ which gives

$$k_{12}\phi_2 = Q - (k_{13} + k_{14})d_a. \tag{7.9.14}$$

This combination of a parabola approximation of the saturated layer with reverse Lynch is not dependent on having $\Delta x = \Delta z$ in the bottom element but of course this is more accurate.

Table 7.9.2 shows results from some numerical examples. Reverse Lynch on its own is limited to mesh 1 where $\Delta x = \Delta z$.

The effect of using the D–F parabola plus reverse Lynch to give the starting values of the hydraulic potential ϕ has been used in some examples. Taking the slope with the dimensions shown in Fig. 7.7.2, $Q = 0.1$ m^3 per hour, $K_\mathrm{s} = 0.5$ m per hour:

1. If the local hydraulic gradient is set to zero in the unsaturated zone the result shows in the run-in a saturated wedge maintained at the foot of the slope. When rainfall starts conditions change rapidly to give a saturated layer.

2. If no-flow conditions are intially imposed in the unsaturated zone above the D–F parabola the upper slopes start off very dry.

A compromise is to take the D–F parabola as indicating the position

Table 7.9.2 Initial saturated zone estimations

	Height of saturated region at $x = \Delta x$ (m)	ϕ_2 (m)	ϕ_3 (m)	ϕ_4 (m)
Example (b)				
$Q = 0.010\,63$ m^3 h^{-1}				
Mesh 1, $K_s = 0.1$ m h^{-1}*				
Model	0.21	0.16	0.20	0.21
Reverse Lynch	0.24		0.24	0.24
D–F parabola	0.33	−0.25	0.33	0.33
D–F adjusted	0.23	0.03	0.23	0.23
Example (c)				
$Q = 0.015\,36$ m^3 h^{-1}				
Mesh 8, $K_s = 0.1$ m h^{-1}, $\psi_{in} = -0.1$ m				
Model	1.17	0.027	1.34	1.17
D–F parabola	1.24	0.027	1.24	1.24
D–F adjusted	1.14	0.027	1.14	1.14
Example (d)				
$Q = 0.023\,43$ m^3 h^{-1}				
Mesh 8, $K_s = 0.5$ m h^{-1}, $\psi_{in} = -0.5$ m				
Model	0.61	0.0085	0.61	0.61
D–F parabola	0.68	0.0095	0.68	0.68
D–F adjusted	0.58	0.0091	0.58	0.58

* Initial ψ values one-tenth of altitude.
These model results are each after a run of 100 hours with no rain.

of the saturated layer near the foot of the slope and gradually decrease the values of ϕ towards the top of the slope. This can be done so as to influence the results more towards those labelled as the u values or those labelled as the v values in Figs 7.9.1 and 7.9.2.

7.10 Three-dimensional groundwater flow

A three-dimensional model of saturated–unsaturated subsurface flow is a natural extension of the two dimensions in the vertical variable breadth model described in section 7.7. But (a) the labour involved in laying out a three-dimensional grid of elements and (b) the computational cost of solving the simultaneous equations are enormously greater. Bricks and prisms seem to be the easiest elements to use to fill the space. Frind and Verge (1978) use isoparametric bricks in which any side can have either linear, quadratic, or cubic variation. The hydraulic conductivity tensor is taken as diagonal and aligned with the global axes and a backward difference scheme is used for the time stepping. The equations are solved by direct Gauss or Cholesky schemes adapted for sparse matrices (Duff *et al.*, 1989). Huyakorn *et al.* (1986) use bricks and prisms and a vertical slice SOR (Successive Over Relaxation) scheme (Appendix 2) to solve the equations. Recent work on PCG (Preconditioned Conjugate Gradient) schemes (Appendix 2) aims to improve cost efficiency in these large problems. Chapter 10 gives examples of more recent work on three-dimensional groundwater flow in connection with water-quality modelling.

8 Boundary integral methods

8.1 Introduction

Boundary integral methods are taken here to mean methods where all or part of the computation is transferred to an integral on the boundary of the region. There is the boundary integral method which is referred to here as the BIM. This is also sometimes called the boundary element method (Latinopoulos, 1986; Grilli, 1989; Latinopoulos and Katsifarakis, 1991) or the boundary integral equation method (Liggett and Liu, 1983). The BIM is described in section 8.2. Sections 8.3 and 8.4 describe two different kinds of problem where boundary integrals are used together with finite elements to cope with regions which are sufficiently large so as to be called infinite.

8.2 The boundary integral method

The BIM is simple to use when the equation to be solved is Laplace's equation

$$\nabla^2 \phi = 0 \qquad (8.2.1)$$

and the boundary of the region R in which the equation is to be solved is known. For groundwater flow this means essentially the two dimensions in plan-vertically-integrated aquifer problem represented by the steady-state form of eqn (1.7.10) with constant transmissivity T throughout the region and no sources or sinks. These can be added in as explained later in this section provided the problem is linear.

There are two basic steps in the BIM as follows:

Step 1: N_n points are selected as 'nodes' on the boundary of the region as shown in Fig. 8.2.1. Each of these belongs either to S_1 where the potential ϕ is known or to S_2 where the flow and hence $\partial\phi/\partial n$ is known. A set of equations is formed from which ϕ is obtained for the nodes on S_2 and $\partial\phi/\partial n$ is obtained for nodes on S_1.

Step 2: From another equation, using these nodal values, it is then possible to obtain the value of ϕ at any point in the interior of the region.

In each of these steps the result requires the use of Green's second identity which is obtained for the two-dimensional case as follows.

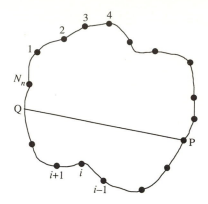

Figure 8.2.1: BIM nodes on boundary.

If U, V are any two functions which can be differentiated twice in the region R, then

$$\int_R (\nabla U \cdot \nabla V + U \nabla^2 V)\, \mathrm{d}R = \int_R \nabla(U \nabla V)\, \mathrm{d}R = \int_S U \frac{\partial V}{\partial n}\, \mathrm{d}S \quad (8.2.2)$$

where $\partial V/\partial n = \nabla V \cdot \mathbf{n}$ and $\mathbf{n}$ is the outward normal; $S = S_1 + S_2$ is the boundary of R. Similarly

$$\int_R (\nabla U \cdot \nabla V + V \nabla^2 U)\, \mathrm{d}R = \int_S V \frac{\partial U}{\partial n}\, \mathrm{d}S. \quad (8.2.3)$$

Subtracting (8.2.3) from (8.2.2) then gives

$$\int_R (U \nabla^2 V - V \nabla^2 U)\, \mathrm{d}R = \int_S \left(U \frac{\partial V}{\partial n} - V \frac{\partial U}{\partial n} \right) \mathrm{d}S. \quad (8.2.4)$$

U is now replaced by the potential ϕ which satisfies $\nabla^2 \phi = 0$ (this is where it matters that the equation for the flow is Laplace's equation) and V is replaced by Green's function

$$V = \ln(r) \quad (8.2.5)$$

which satisfies Laplace's equation everywhere in R except at $r = 0$ where it has a singularity. Because of the logarithm here it is essential that lengths are dimensionless. If $\hat{r}$ is the actual distance, then the value of r is $\hat{r}/L$ where L is a 'diameter' of the region R. Then eqn (8.2.4) reduces to

$$\int_S \left(\phi \frac{\partial V}{\partial n} - V \frac{\partial \phi}{\partial n} \right) \mathrm{d}S = 0. \quad (8.2.6)$$

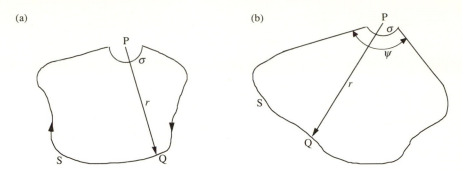

Figure 8.2.2: (a) Smooth boundary at P; (b) angle ψ at P.

Suppose P is a point on the boundary of R. P may be a point where the boundary is smooth as in Fig. 8.2.2(a) or where there is a corner of internal angle ψ as in Fig. 8.2.2(b). P is taken to be the point where $r = 0$, i.e. where Green's function $\ln(r)$ has a singularity (this property is sometimes described as 'going to infinity' but this is not a precise definition and strictly the function does not exist at this point which is what 'having a singularity' means). Because of the singularity at P the boundary S is taken to avoid P by enclosing it in an arc σ of a circle centre P, radius $r = r_0$; then r_0 is allowed to tend to zero.

Substituting into eqn (8.2.6) gives

$$\int_S \left(\frac{\phi}{r} \frac{\partial r}{\partial n} - \ln(r) \frac{\partial \phi}{\partial n} \right) dS + \lim_{r_0 \to 0} \int_\sigma \left(\frac{\phi}{r_0} \frac{\partial r}{\partial n} - \ln(r_0) \frac{\partial \phi}{\partial n} \right) r_0 \, d\theta = 0. \quad (8.2.7)$$

Now on the circle σ, the outward normal to R is inward towards the centre P of the circle; hence $\partial r/\partial n = -1$ and the second integral on the left-hand side of (8.2.7) becomes

$$I_2 = \lim_{r_0 \to 0} \left[\int_\sigma -\phi \, d\theta + r_0 \ln(r_0) \int_\sigma \frac{\partial \phi}{\partial r} \, d\theta \right]$$

$$= -\psi\phi(\mathrm{P}) + \lim_{r_0 \to 0} r_0 \ln(r_0) \int_\sigma \frac{\partial \phi}{\partial r} \, d\theta.$$

But $\partial\phi/\partial r$ is bounded on σ and $\lim_{r_0 \to 0} r_0 \ln(r_0)$ is zero. Hence

$$I_2 = -\psi\phi(\mathrm{P}) \qquad (8.2.8)$$

where ψ is the angle subtended by the tangents to the boundary on either side of P. If the curve here is smooth as in Fig. 8.2.2(a) then $\psi = \pi$.

Substituting (8.2.8) for the second integral in (8.2.7) gives

$$\psi\phi(P) = \int_S \left(\frac{\phi}{r}\frac{\partial r}{\partial n} - \ln(r)\frac{\partial\phi}{\partial n}\right)dS. \tag{8.2.9}$$

By choosing the point P to be at the nodes P_j, $j = 1, 2, \ldots, N_n$, in turn and approximating the integral numerically, N_n equations are formed to give the N_n unknowns required to be found in step 1. These are the values of ϕ at nodes on S_2 and values of $\partial\phi/\partial n$ at nodes on S_1. The integral in (8.2.9) is approximated as a sum

$$\psi\phi(P_i) = \sum_j I_{i,j} \tag{8.2.10}$$

where $I_{i,j}$ is the integral over the segment Q_jQ_{j+1} shown in Fig. 8.2.3. Q_jQ_{j+1} is taken to be a straight line and if η_i is the perpendicular distance from P_i to this line and ξ is the distance of a general point Q on this line from the foot of the perpendicular as shown, then at Q

$$\frac{\partial r}{\partial n} = \frac{\partial r}{\partial \eta_i} = \frac{\partial}{\partial \eta_i}\sqrt{(\eta_i^2 + \xi^2)} = \frac{\eta_i}{r}. \tag{8.2.11}$$

The two simplest approximations for the integrals are (a) the midpoint value times the length of segment:

$$I_{i,j} = \left[\frac{\phi(Q_{j+1/2})}{r_{i,j+1/2}^2}\eta_i - \ln(r_{i,j+1/2})\frac{\partial}{\partial n}\phi(Q_{j+1/2})\right] \text{ times } Q_jQ_{j+1} \tag{8.2.12a}$$

or (b) the trapezium rule, i.e.

$$I_{i,j} = \left[\frac{1}{2}\left(\phi(Q_j)\frac{\eta_i}{r_{i,j}^2} + \phi(Q_{j+1})\frac{\eta_i}{r_{i,j+1}^2}\right)\right.$$
$$\left. - \frac{1}{2}\left(\ln(r_{i,j})\frac{\partial}{\partial n}\phi(Q_j) + \ln(r_{i,j+1})\frac{\partial}{\partial n}\phi(Q_{j+1})\right)\right] \text{ times } Q_jQ_{j+1}$$
$$\tag{8.2.12b}$$

with all lengths scaled with respect to L. Substituting into eqn (8.2.10) and repeating for $i = 1, 2, \ldots, N_n$ gives the set of N_n simultaneous linear equations to complete step 1. Higher-order approximations for the integrals are given by Liggett and Liu (1982).

For step 2 the singularity point P is now taken as an interior point and surrounded by a complete circle σ joined to S as shown in Fig. 8.2.4. Since the integrals along the lines connecting σ to S cancel the result is

$$2\pi\phi(P) = \int_S \left(\frac{\phi}{r}\frac{\partial r}{\partial n} - \ln(r)\frac{\partial\phi}{\partial n}\right)dS. \tag{8.2.13}$$

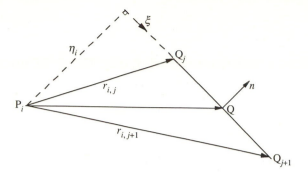

Figure 8.2.3: The boundary integral over $Q_j Q_{j+1}$.

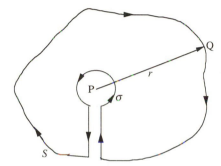

Figure 8.2.4: An interior point P.

With the values of ϕ and the normal derivatives all known at the nodes the values of $\phi(P)$ can be calculated at any point in the interior with the same approximation as in step 1. However, the accuracy deteriorates if the point P is too near the boundary (a disadvantage which does not occur with the finite element method). Liggett and Liu (1983) give a 'rule of thumb' which says that it is better to take the point P at a distance from the boundary segment $Q_j Q_{j+1}$ greater than the length of this segment.

The number of nodes on the boundary can be doubled and the process repeated until the difference between successive solutions is less than some given tolerance.

Source or sink terms in a confined aquifer

These are included by superposition, i.e. $\phi(P_i)$ is replaced by $\phi(P_i) + \phi_{iw}$ where

$$\phi_{iw} = \frac{Q_w}{2\pi T} \ln(r_{iw}) \tag{8.2.14}$$

(see the analytic solution, eqn 1.8.11). Here Q_w is the strength of the source and r_{iw} is the (dimensionless) distance from the source. Thus in solving for $\phi(P_i)$ without the source the boundary conditions in step 1 must be adjusted to allow for the ϕ_w part of the solution, i.e. if the boundary conditions are given as

$$\phi = \alpha \text{ (known head) on } S_1 \quad \text{and} \quad \frac{\partial \phi}{\partial n} = \gamma \text{ (known flow) on } S_2$$

then in solving for $\phi(P_i)$ they are adjusted to

$$\phi(P_i) = \alpha - \frac{Q_w}{2\pi T} \ln(r_{iw}) \text{ on } S_1$$

and

$$\frac{\partial \phi}{\partial n}(P_i) = \gamma - \frac{Q_w}{2\pi T r_{iw}} \frac{\partial r}{\partial n} \text{ on } S_2.$$

Then the original boundary conditions are satisfied by $\phi(P_i) + \phi_{iw}$. This method can be extended to deal with a region with many abstraction and recharge wells.

Latinopoulos (1986) uses a boundary integral method for the accurate estimate of velocities in the groundwater near wells in an isotropic confined aquifer. The components of the velocity are needed for the solution of the advection–diffusion equation in water-quality modelling —see Chapter 9. With approximation (a) for the integrals and with the addition of the wells eqn (8.2.13) becomes

$$2\pi\phi(P_i) = -\sum_w \frac{Q_w}{T} \ln(r_{iw}) + \sum_j \phi_{j+1/2} \int \frac{1}{r_{i,j}} \frac{\partial r_{i,j}}{\partial n} dS_j$$

$$- \sum \frac{\partial \phi_{j+1/2}}{\partial n} \int \ln(r_{i,j}) dS_j. \tag{8.2.15}$$

The velocity components

$$v_x = -T \frac{\partial \phi}{\partial x}(P_i), \qquad v_y = -T \frac{\partial \phi}{\partial y}(P_i)$$

are then obtained by differentiating (8.2.15) and performing the integration analytically.

Latinopoulos and Katsifarakis (1991) give a method for dealing with an aquifer divided into M zones each with a constant transmissivity

T_m, $m = 1, 2, \ldots, M$. Step 1 is performed for each zone separately. Then for each node j on the boundary between zone m and zone $m + 1$, say, there are compatibility conditions

$$\phi_{j,m} = \phi_{j,m+1} \tag{8.2.16}$$

and

$$T_m\left(\frac{\partial\phi}{\partial n}\right)_{j,m} = -T_{m+1}\left(\frac{\partial\phi}{\partial n}\right)_{j,m+1} \tag{8.2.17}$$

Suppose for each zone step 1 gives a set of equations

$$\mathbf{A}_m\{\phi\} = \mathbf{B}_m\{\partial\phi/\partial n\}. \tag{8.2.18}$$

Then the M such sets of equations are combined with the compatibility conditions (8.2.16) and (8.2.17) to form a single set of equations to be solved for the unknowns on the boundary of the whole region plus the values at the nodes on the interzone boundaries.

Shapiro and Andersson (1983) used a coupled model for the steady-state flow in fractured porous rock. The BIM is used for the fluid flow in the rock using the head and the fluid mass flux at the boundaries. Then one-dimensional equations for the flow in the fractures are expressed in terms of the same variables. Elsworth (1987) uses the BIM (in the linear region) and the FEM (in the non-linear region) for porous and fractured media flow. These methods depend on knowing the location of the fractures. Berkowitz *et al.* (1988) examine the conditions under which an equivalent single continuum porous medium can be used in a study of contaminant transport in fractured porous formations. This depends on obtaining an equivalent porosity and coefficient of dispersion from field tests.

The BIM can be used to solve the two-dimensions-in-elevation problem for the position of the phreatic surface S_s (section 1.7). On this surface there are the two boundary conditions

$$\text{(a)} \quad \frac{\partial\phi}{\partial n} = 0 \quad \text{and} \quad \text{(b)} \quad \phi = z. \tag{8.2.19}$$

Suppose the height of a node on the phreatic surface is assumed at the start to be z^0 and the value for ϕ obtained here at the first application of the BIM is ϕ^1. Then instead of taking the position of the node at the next iteration as $z^1 = \phi^1$, it is better to take an average

$$z^1 = 0.5(\phi^1 + z^0) \tag{8.2.20}$$

(as in section 7.6) and continue to average in this way to avoid oscillations.

Grilli (1989) estimates the ratio of the CPU times needed to solve a problem by the BIM and the finite element method. Suppose the BIM uses N_B nodes and the FEM uses N_F nodes. Then the BIM needs the solution of a full matrix with a number of operations proportional to N_B^3 and the FEM needs the solution of a sparse banded matrix with a number of operations proportional to N_F^2 (Conte and de Boor, 1980). Supposing that $N_F = 30 N_B$, $100 N_B$ for comparable accuracy in the solution of two- and three-dimensional problems respectively, the ratio,

$$ r = \frac{N_B^3}{N_F^2} = \begin{cases} N_B/900 & \text{in two dimensions} \\ N_B/10\,000 & \text{in three dimensions} \end{cases}. \qquad (8.2.21) $$

Hence with $N_B = 100$ in a two-dimensional problem the solution time is nine times faster with the BIM. With $N_B = 1000$ in a three-dimensional problem the BIM is 10 times faster. For a very large problem where $r > 1$ Grilli suggests a subregion division of the computation to make sure the BIM is faster.

8.3 Groundwater flow round a tunnel with permeable invert

In sections 8.3 and 8.4 two different examples of exterior boundary value problems are discussed. These are problems where the region in which the problem is to be solved is exterior to some object and extends to some indeterminate large distance. Usually with this kind of problem it is the near-field solution which matters. It is possible to use finite elements only, repeating the solution with the region increasing in size until the difference between successive solutions is less than a given tolerance. But in some circumstances it is more cost-effective to combine the finite elements in an inner region R_i with a boundary integral method which extends the solution out to an infinite distance. The example used here is an exterior boundary value problem of the groundwater flow round a circular tunnel which runs under a river. The tunnel has a permeable invert, i.e. the water is allowed to enter at the bottom of the tunnel and is pumped away. The radius of the tunnel is r_i; the river is of depth d and is supposed to be sufficiently wide so that near the centre the flow can be taken as two dimensional in the vertical section. The ground is saturated up to the bottom of the river. Supposing the ground is isotropic and the hydraulic conductivity is constant, the hydraulic potential satisfies Laplace's equation (8.2.1). The flow is symmetric about the vertical through the centre of the tunnel so it is only necessary to solve over half the region as shown in Fig. 8.3.1 (not to scale). The region R over which the

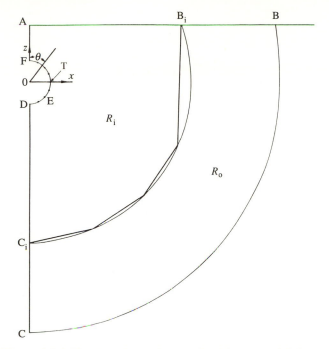

Figure 8.3.1: Section through tunnel with permeable invert.

problem is to be solved is the region exterior to the tunnel bounded above by the bottom of the river and stretching out to some large distance which can be taken as infinite. With axes $0x$, $0z$ as shown the boundary conditions are as follows:

(1) $\phi = a + d$ on AB, where a is the depth of the centre of the tunnel below the bottom of the river;

(2) $\phi = z$ on DE, i.e. $r = r_t$, $\theta_1 < \theta \leq \pi$, the permeable invert of the tunnel which is open to the atmosphere where the pressure is taken as zero;

(2) $\partial \phi / \partial x = 0$ on CD and FA from symmetry;

(4) $\partial \phi / \partial r = 0$ on the arc of the tunnel EF, i.e. $r = r_t$, $0 \leq \theta \leq \theta_1$; this is a no-flow condition where the tunnel lining is impermeable.

At a large distance from the tunnel it is supposed that there is no flow and to conform with the value on AB the solution here is taken as

$$\phi_\infty = a + d. \tag{8.3.1}$$

The problem is simplified by subtracting out the solution at infinity, i.e. put

$$\phi' = \phi - \phi_\infty. \tag{8.3.2}$$

Then

$$\nabla^2 \phi' = 0 \tag{8.3.3}$$

and going round clockwise starting from A the boundary conditions on ϕ' are

$$\phi' = 0 \text{ on AB} \tag{8.3.4a}$$

$$\phi' = 0 \text{ on the arc BC, at an infinite distance} \tag{8.3.4b}$$

$$\frac{\partial \phi'}{\partial x} = 0 \text{ on CD (symmetry)} \tag{8.3.4c}$$

$$\phi' = r_{\rm t} \cos\theta - (a + d), \qquad \theta_1 < \theta \le \pi, \text{ on DE} \tag{8.3.4d}$$

$$\frac{\partial \phi'}{\partial r} = 0, \qquad 0 \le \theta \le \theta_1, \text{ on EF} \tag{8.3.4e}$$

$$\frac{\partial \phi'}{\partial x} = 0 \text{ on FA (symmetry).} \tag{8.3.4f}$$

The problem is solved numerically by taking linear triangle finite elements in the interior region $R_{\rm i}$ in Fig. 8.3.1 bounded by AB$_{\rm i}$C$_{\rm i}$DEFA (where B$_{\rm i}$C$_{\rm i}$ is the polygonal boundary with nodes on a circle concentric with the tunnel and of radius $r_{\rm i}$) with basis functions $N_j = N_j(x, y)$ and combining these with a parameter b times a function $\hat\phi$ which exists only in the region R_0 external to $R_{\rm i}$, i.e.

$$\phi' = \sum_j \phi'_j N_j \text{ (in } R_{\rm i}) + b\hat\phi \text{ (in } R_0) \tag{8.3.5}$$

where

$$\hat\phi = \frac{\cos\theta}{r} - \frac{2a - r\cos\theta}{r^2 + 4a^2 - 4ar\cos\theta}.$$

This function $\hat\phi$ is the solution of Laplace's equation which represents a source at $r = 0$ together with its image at $x = 0$, $z = 2a$; these combine to give $\hat\phi = 0$ on B$_{\rm i}$B and $\partial\hat\phi/\partial x = 0$ on CC$_{\rm i}$; also $\hat\phi$ tends to zero as r tends to infinity. The solutions in $R_{\rm i}$ and R_0 are to be joined with as much continuity as possible.

Equation (8.3.3) is first multiplied by a test function w and integrated by parts over the inner region $R_{\rm i}$ only to obtain the weak form:

$$\int_{R_1} \nabla w \cdot \nabla \phi' \, \mathrm{d}R = \int_{S_1} w \frac{\partial \phi'}{\partial n} \, \mathrm{d}S \tag{8.3.6}$$

where S_i is the complete boundary of R_i, i.e. AB_iC_iDEFA. Substituting from (8.3.5) gives

$$\sum_j \phi'_j \int_{R_i} \nabla w \cdot \nabla N_j \, dR = \int_{S_i} w \frac{\partial \phi'}{\partial n} \, dS. \tag{8.3.7}$$

Then for $w = N_k$, a basis function,

$$\sum_j \phi'_j \int_{R_i} \nabla N_k \cdot \nabla N_j \, dR = \int_{AB_i} N_k \frac{\partial \phi'}{\partial n} \, dz + \int_{B_iC_i} N_k \frac{\partial \phi'}{\partial n} \, dS$$

$$+ \int_{DE} N_k \frac{\partial \phi'}{\partial n} \, dS - r_t \int_{EF} N_k \cos \theta \, d\theta. \tag{8.3.8}$$

When k is a node on AB_i or on DE the values of ϕ' are known from (8.3.4a and d) and hence these equations are to be overwritten. The boundary condition $\partial \phi'/\partial n = b \, \partial \hat\phi/\partial n$ is inserted into the integral along B_iC_i, i.e. this becomes

$$\int_{B_iC_i} b N_k \frac{\partial \hat\phi}{\partial n} \, dS.$$

Thus this set of simultaneous linear equations can be written in the form

$$\mathbf{K}_{11}\phi'_i + b\mathbf{K}_{12} = \mathbf{f}_1 \tag{8.3.9}$$

where $\mathbf{K}_{11}$ is the standard finite element stiffness matrix, ϕ'_i is the vector of unknown nodal values in R_i, and the kth entry in $\mathbf{K}_{12}$ is

$$\int_{B_iC_i} N_k \frac{\partial \hat\phi}{\partial n} \, dS. \tag{8.3.10}$$

The normal here is in the direction outwards from R_i. The vector $\mathbf{f}_1$ contains the terms from the essential boundary condition (8.3.4d).

Next, using the weak form of the equation

$$\nabla^2 b\hat\phi = 0$$

with the test function $w = \hat\phi$ and integrating over the outer region R_0 only, gives

$$\int_{R_0} b \nabla \hat\phi \cdot \nabla \hat\phi \, dR = \int_{S_0} b\hat\phi \frac{\partial \hat\phi}{\partial n} \, dS. \tag{8.3.11}$$

The left-hand side is replaced by the boundary integral and on the right-hand side the boundary conditions to be satisfied are written in.

Table 8.3.1

r_i		49.5	69.0	96.1	134.0	186.0
T	(a)	39.56	38.74	38.28	38.01	37.86
	(b)	37.64	37.62	37.64	37.65	37.66
F	(a)	58.58	57.71	57.21	56.93	56.76
	(b)	56.55	56.51	56.53	56.55	56.56

The only non-zero integrand is that along $B_i C_i$. Thus the result is

$$\int_{B_i C_i} \left(b\hat{\phi} - \sum N_j \phi_j' \right) \frac{\partial \hat{\phi}}{\partial r} \, dS = 0. \tag{8.3.12}$$

(This is essentially a weak satisfaction of the continuity condition $b\hat{\phi} = \sum N_j \phi_j'$ on $B_i C_i$). Equation (8.3.12) can be written in the form

$$\mathbf{K}_{12}^{T} \boldsymbol{\phi}_i' - b K_{11} = 0 \tag{8.3.13}$$

where

$$K_{11} = \int_{C_i B_i} \hat{\phi} \frac{\partial \hat{\phi}}{\partial r} \, dS$$

which can be found numerically by the trapezium rule for example as described in section 8.2. Thus if there are N_n unknown nodal values in R_i eqns (8.3.9) and (8.3.13) provide $N_n + 1$ equations to solve for these nodal values and the parameter b. The solution near the tunnel is of most interest and the boundary $B_i C_i$ can be moved outwards until the difference between successive solution values is less than some tolerance. Table 8.3.1 (adapted from Wood, 1976) compares the convergence with (a) the finite element solution only in the inner region radius r_i, taking $\phi = a + d$ on the polygonal boundary $B_i C_i$, and (b) the finite element solution in the inner region coupled to the boundary integral solution in the infinite region. Table 8.3.1 shows values of ϕ at points T, $(r_t, 0)$ and F, $(0, r_t)$ on the surface EF. These values are calculated with $d = 60$, $a = 30$, $r_t = 3.5$, and the arc DE subtending $45°$ at the centre of the tunnel. The results show the improved convergence with method (b).

8.4 Electromagnetic river gauging (Trkov and Wood, 1980)

This is a three-dimensional example of an exterior boundary value problem solved by using a combination of finite elements together with infinite elements. A magnetic field is induced by energizing coils placed under the river bed. A potential difference is established between two probes placed on opposite banks of the river. The magnitude of the induced potential is related to the flow of water by the formula

$$\text{potential difference} = \int_{\text{river}} \mathbf{v} \cdot (\mathbf{B} \wedge \nabla\phi) \, dR \qquad (8.4.1)$$

where $\mathbf{v}$ is the velocity in the river, $\mathbf{B}$ is the magnetic field, and ϕ is the potential due to unit flux between the probes, (Ferraro, 1962). The problem considered here is that of solving for the potential ϕ. This is normalized to be ± 1 on the probes and ϕ also satisfies the equation

$$\nabla(\sigma\nabla\phi) = 0 \qquad (8.4.2)$$

(where σ is the conductivity) in the three-dimensional region surrounding the probes which includes the river and the ground. The values of the conductivity are taken as $\sigma = 0$ in the air, $\sigma = \sigma_r$ (constant) in the river, and $\sigma = \sigma_g$ (constant) in the ground. The conductivity could be taken as variable in the part of the region represented by finite elements but not in the infinite elements; hence it is simpler to assume it just has these constant average values. The river is assumed to flow in a straight bed of constant trapezoidal section as shown in Fig. 8.4.1. The river is assumed to be full. This would not be necessary with finite elements only but it is necessary for the assumptions made for the boundary integral solutions in the infinite elements. The vertical plane

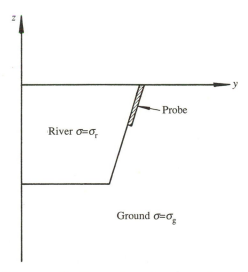

Figure 8.4.1: Vertical section through probe.

through the probes is taken as $x = 0$ and the y and z axes form a right-handed set as in Fig. 8.4.1. The probes are assumed to be linear and placed on the river–ground interface.

Thus the problem to be solved for the potential ϕ is as follows: $\phi(x, y, z)$ satisfies eqn (8.4.2) with boundary conditions

$$\phi = \pm 1 \text{ on the probes} \tag{8.4.3a}$$

$$\text{on } y = 0, \; \phi(x, 0, z) = 0, \text{ from symmetry} \tag{8.4.3b}$$

$$\text{on } x = 0, \; \partial\phi/\partial n = 0, \text{ from symmetry} \tag{8.4.3c}$$

$$\text{because } \sigma = 0 \text{ in the air, } \partial\phi/\partial n = 0 \text{ on } z = 0 \tag{8.4.3d}$$

$$\text{at an infinite distance the potential } \phi \text{ is zero.} \tag{8.4.3e}$$

At the interface S_{rg} between the river and the ground there are two boundary conditions:

$$\phi(\text{river}) = \phi(\text{ground}) \tag{8.4.3f}$$

$$\sigma_r \frac{\partial\phi}{\partial n} (\text{river}) = \sigma_g \frac{\partial\phi}{\partial n} (\text{ground}) \tag{8.4.3g}$$

where the derivatives are taken to be in the direction of the normal to the river–ground interface drawn from the river to the ground.

Because of the symmetry the problem need only be solved in one octant of the three-dimensional space, and hence only the region $x \geq 0$, $y \geq 0$, $z \geq 0$ is now considered.

The method used for solution of this problem is an extension of the method used for the two-dimensional groundwater flow round a tunnel problem in section 8.3. In a box of finite elements near the probe the solution is taken to be

$$\sum_j \phi_j N_j(x, y, z) \tag{8.4.4}$$

where the elements are eight-node right prisms with quadrilateral bases parallel to the y–z plane as shown in Fig. 8.4.2. The box has edges parallel to the axes. This box of finite elements is joined to two infinite elements in the regions R_r(river) and R_g(ground) which fill the remainder of the octant. The conditions (8.4.3f) and (8.4.3g) on the boundary S_{rg} have to be included to give as much continuity as possible on the boundaries S_{ir} and S_{ig} between the box and the infinite elements in river and ground respectively. In the infinite elements it is convenient to take the solution as a parameter times

$$\phi_\infty = yr^{-3/2} \tag{8.4.5}$$

(a)

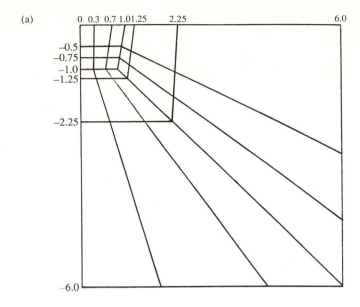

(b)

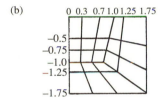

Figure 8.4.2: Sections through box elements.

where r is the distance from the origin, i.e. $r^2 = x^2 + y^2 + z^2$. The function (8.4.5) is the simplest solution of Laplace's equation which also satisfies conditions (8.4.3b–e). The approximate solution of the problem is thus expressed in the form

$$\phi \approx \sum_j \phi_j N_j \text{ (in the box)} + a\phi_\infty \text{ (in the rest of the river)}$$

$$+ b\phi_\infty \text{ (in the rest of the ground)}.$$

The method described here is essentially the same as that used in section 8.3.

First the weak form of eqn (8.4.2) with the test function equal to a basis function, N_k, is used in the inner box only with the boundary conditions to be written in on the right-hand side:

$$\sum_j \phi_j \int_{R_i} \sigma \nabla N_k \cdot \nabla N_j \, dR = \int_{S_b} \sigma N_k \frac{\partial \phi}{\partial n} \, dS$$

(where S_b is the surface of the box)

$$= a \int_{S_{ir}} \sigma_r N_k \frac{\partial \phi_\infty}{\partial n} \, dS + b \int_{S_{ig}} \sigma_g N_k \frac{\partial \phi_\infty}{\partial n} \, dS. \qquad (8.4.6)$$

The essential boundary conditions, $\phi = 0$ on $y = 0$ and $\phi = 1$ on the probe, are overwritten; hence the number of equations in (8.4.6) is equal to the number of unknown nodal values ϕ_j. Equation (8.4.6) can be written in the form

$$\mathbf{K}_{11}\boldsymbol{\phi} - a\sigma_r\mathbf{K}_{12} - b\sigma_g\mathbf{K}_{13} = \mathbf{f}_1 \qquad (8.4.7)$$

where $\mathbf{K}_{11}$ is the usual stiffness matrix, $\mathbf{K}_{12}$ and $\mathbf{K}_{13}$ are vectors; the vector $\mathbf{f}_1$ contains the terms from the boundary condition (8.4.3a).

The second equation is obtained by taking the weak form of eqn (8.4.2) in the region R_r, i.e. the river external to the box, taking the test function equal to ϕ_∞, and writing in the boundary conditions on the right-hand side. The result is

$$\int_{R_r} a\sigma_r \nabla\phi_\infty \cdot \nabla\phi_\infty \, dR = \int_{S_{ir}} \sigma_r \sum_j N_j \phi_j \frac{\partial \phi_\infty}{\partial n} \, dS + \int_{S_{rg}} b\sigma_g \phi_\infty \frac{\partial \phi_\infty}{\partial n} \, dS. \qquad (8.4.8)$$

Since the left-hand side of (8.4.8) is equal to

$$\int_{S_{ir}+S_{rg}} a\sigma_r \phi_\infty \frac{\partial \phi_\infty}{\partial n} \, dS$$

(8.4.8) is equivalent to a weak satisfaction of the boundary conditions on S_{ir} and S_{rg}:

$$\int_{S_{ir}} \sigma_r \left(\sum_j N_j \phi_j - a\phi_\infty \right) \frac{\partial \phi_\infty}{\partial n} \, dS - \int_{S_{rg}} (a\sigma_r - b\sigma_g)\phi_\infty \frac{\partial \phi_\infty}{\partial n} \, dS = 0. \qquad (8.4.9)$$

Equation (8.4.9) can be written as

$$\sigma_r \mathbf{K}_{12}^T \boldsymbol{\phi} - a\sigma_r \left[\int_{S_{ir}} \phi_\infty \frac{\partial \phi_\infty}{\partial n} \, dS + \int_{S_{rg}} \phi_\infty \frac{\partial \phi_\infty}{\partial n} \, dS \right] + b\sigma_g \int_{S_{rg}} \phi_\infty \frac{\partial \phi_\infty}{\partial n} \, dS = 0.$$
$$(8.4.10)$$

Similarly the third equation is a weak satisfaction of the boundary conditions on S_{ig} and S_{rg} obtained from the weak form in the ground external to the box. The result is

$$\sigma_g \mathbf{K}_{13}^T \boldsymbol{\phi} + a\sigma_r \int_{S_{rg}} \phi_\infty \frac{\partial \phi_\infty}{\partial n} \, dS$$

$$+ b\sigma_g \left[-\int_{S_{ig}} \phi_\infty \frac{\partial \phi_\infty}{\partial n} \, dS - \int_{S_{rg}} \phi_\infty \frac{\partial \phi_\infty}{\partial n} \, dS \right] = 0 \qquad (8.4.11)$$

where the signs have been adjusted so that the normal derivative is always from the river towards the ground.

Substituting

$$A = \int_{S_{ir}} \phi_\infty \frac{\partial \phi_\infty}{\partial n} \, dS, \quad B = \int_{S_{ig}} \phi_\infty \frac{\partial \phi_\infty}{\partial n} \, dS, \quad C = \int_{S_{rg}} \phi_\infty \frac{\partial \phi_\infty}{\partial n} \, dS$$

eqns (8.4.7), (8.4.10), and (8.4.11) can be written as the system of equations

$$\left. \begin{array}{l} \mathbf{K}_{11}\boldsymbol{\phi} - a\sigma_r \mathbf{K}_{12} - b\sigma_g \mathbf{K}_{13} = \mathbf{f}_1 \\[2mm] \sigma_r \mathbf{K}_{12}^T \boldsymbol{\phi} - a\sigma_r(A + C) + b\sigma_g C = 0 \\[2mm] \sigma_g \mathbf{K}_{13}^T \boldsymbol{\phi} + a\sigma_r C + b\sigma_g(-B - C) = 0. \end{array} \right\} \qquad (8.4.12)$$

This system of equations can be solved for the unknown nodal values ϕ_j and the parameters a and b. Trkov and Wood (1980) (using a variational approach) show that it is more cost-effective to use a composite method with finite elements in an inner box (section shown in Fig. 8.4.2(a)) plus the two infinite elements than to approximate the region only with finite elements as shown in section in Fig. 8.4.2(b). For the latter 185 nodes are required whereas the composite method requires only 120 nodes to produce almost the same results. This result is obtained using eight-node elements which are right prisms with quadrilateral bases parallel to the y–z plane. The stiffness matrix is calculated using $2 \times 2 \times 2$ Gauss numerical integration. The surface S_{rg} is infinite in the x direction and this part of the integral on this surface is evaluated exactly. In the other direction Simpson's rule is used; the number of subintervals can then easily be doubled to check the accuracy.

9 Water-quality modelling

9.1 Introduction

Water-quality modelling is a very active research area and there are two distinct trends towards:

(1) parallel algorithms to handle larger and larger problems;

(2) adaptations of algorithms to be run on personal computers.

Since this is a complicated subject all that is attempted here is a broad outline of current methods plus references.

The mass-balance equations for a number of chemically interactive solutes of concentrations C_j, $j = 1, 2, \ldots, M$, in the water are each a combination of the physical transport given by the advection and diffusion and the reactions of a particular solute with the other chemicals which are present. They can be written in the form

$$\underset{(1)}{\frac{\partial(nC_j)}{\partial t}} + \underset{(2)}{\mathbf{V} \cdot n[C_j \mathbf{v}} - \underset{(3)}{(D + D^*)\mathbf{V}C_j]} = \underset{(4)}{\sum_{k=1}^{M} S_{jk}(C_1, C_2, \ldots, C_M)} \quad (9.1.1)$$

$j = 1, 2, \ldots, M$, where C_j, $j = 1, 2, \ldots M$, are the concentrations of the substances, $\mathbf{v}$ is the velocity vector, D, D^* are the mechanical (dispersion) and molecular diffusion tensors respectively, and the coupling terms S_{jk} may be non-linear. (Some writers discriminate between molecular diffusion and mechanical dispersion but here the term diffusion is used to cover both.) This equation covers both surface and subsurface flow; the parameter n is the porosity of the ground in subsurface flow and is unity for surface flow.

With Cartesian coordinates and velocity components v_j, the components of the mechanical diffusion are compactly written in the tensor form which will do for two or three dimensions:

$$D_{ij} = a_T v \delta_{ij} + (a_L - a_T) v_i v_j / v \quad (9.1.2)$$

where δ_{ij} is the Kronecker delta ($\delta_{ij} = 1$ if $i = j$ and zero if $i \neq j$), a_L, a_T are the longitudinal and transverse dispersivities, and $v^2 = \Sigma v_j^2 = v_j v_j$ in tensor notation, for example, see Bear (1979). The molecular diffusion term D^* is discussed further for surface flow in section 9.2 and for subsurface flow in section 9.4.

So long as the contaminant is fully dissolved so that the fluid can be assumed to be of uniform density, the continuity and momentum equations are uncoupled from the advection–diffusion equation. From the equations above it can be seen that the velocity components are required for both the advection and diffusion terms. The basic procedure is to obtain the velocity components from the fluid flow equations and substitute them into the advection and diffusion terms. With surface flow the velocity components are directly available from the solution of the continuity and momentum equations. For subsurface flow the problem is more complicated because the components of the velocity must be obtained using the Darcy equation, (1.7.1), from the solution for the piezometric head. Various ways of doing this are discussed in section 9.3.

If there is more than one contaminant and they do not interact then the transport equation (9.1.1) can be solved for each separately using one of the methods described in section 3.8. If there are chemical reactions between the contaminants the equations may be solved as fully coupled or sequentially. There is a spectrum of methods to deal with this situation varying from approximate particle tracking to highly sophisticated methods depending on the computer power available.

The three main solution strategies in the numerical solution of the equations are:

1. Discretize the equations (9.1.1) and solve the whole lot simultaneously by a one-step method (the monolithic approach).

2. Split the differential operator of eqn (9.1.1) into:

(a) The solution of the advection–diffusion part: terms (1) plus (2) plus (3). In order to deal with advection-dominated flow this requires some form of upstream differencing or upwinding. This is because central difference schemes are liable to produce oscillations. Upwinding introduces numerical dispersion but this is often regarded as more acceptable than the possible occurrence of negative values of the concentrations. Upstream weighting with finite elements is referred to in section 9.4.

(b) The solution of the chemical interaction part: terms (1) plus (4) if required. This is usually treated as a system of ordinary differential equations.

3. The Euler–Lagrange approach splits the advection and diffusion terms. The advection terms are treated by the Lagrangian 'follow the particle' along the stream line/characteristic. Interpolation is used to get the values back on to a fixed grid and the diffusion terms are dealt with on this.

Thus the differential operator is split into (a) advection terms (1) plus (2), (b) diffusion terms (1) plus (3), (c) chemical interaction terms (if required) (1) plus (4).

Section 9.2 considers solute transport in surface flow and gives examples of the numerical solution of various problems by the finite difference methods described in Chapter 3. Section 9.4 gives examples of subsurface water-quality models.

Section 9.5 discusses the problem of salt water intrusion in a coastal aquifer. Here the different densities of the fresh and salt water mean that the flow and the advection–diffusion equations are coupled and must be solved together or iterated.

9.2 Surface flow—water-quality modelling

In surface flow if the contaminant is fully dissolved and well mixed so that the water is of uniform density the equations are usually vertically integrated. The vertically integrated two-dimensions-in-plan advection–diffusion equation is

$$\frac{\partial}{\partial t}(HC) + \left[\frac{\partial}{\partial x}(CuH) + \frac{\partial}{\partial y}(CvH)\right] = \frac{\partial}{\partial x}\left(HD_{xx}\frac{\partial C}{\partial x} + HD_{xy}\frac{\partial C}{\partial y}\right)$$

$$+ \frac{\partial}{\partial y}\left(HD_{yx}\frac{\partial C}{\partial x} + HD_{yy}\frac{\partial C}{\partial y}\right) \quad (9.2.1)$$

where H is the depth of the water, C is the depth-averaged concentration of the dissolved contaminant, u, v are the components of the depth-averaged velocity in the x and y directions, and $D_{xx}, D_{xy}, D_{yx}, D_{yy}$ are the depth-averaged diffusion coefficients given by

$$D_{xx} = \frac{(a_L u^2 + a_T v^2)H\sqrt{g}}{(u^2 + v^2)^{1/2}C'} \quad (9.2.2a)$$

$$D_{yy} = \frac{(a_L v^2 + a_T u^2)H\sqrt{g}}{(u^2 + v^2)^{1/2}C'} \quad (9.2.2b)$$

$$D_{xy} = D_{yx} = \frac{(a_L - a_T)uvH\sqrt{g}}{(u^2 + v^2)^{1/2}C'} \quad (9.2.2c)$$

where a_L, a_T are the longitudinal and transverse diffusion coefficients, g is the acceleration due to gravity, and C' is the Chezy bed friction coefficient (section 1.2). These terms can also be written with the Darcy–Weisbach factor, f (section 1.2), using $C' = \sqrt{(8g/f)}$. (Recall that $C'/\sqrt{g}$ is dimensionless.) Source and decay terms can be added to the right-hand side of (9.2.1).

The vertically integrated one-dimensional advection–diffusion equation can be taken as

$$\frac{\partial}{\partial t}(AC) + \frac{\partial}{\partial x}(qC) - \frac{\partial}{\partial x}\left(AD\frac{\partial C}{\partial x}\right) = -FAC + Q_L C_L \qquad (9.2.3)$$

where C is the concentration of the contaminant, A is the cross-sectional area, q is the discharge, D is the diffusion coefficient, F is the linear decay coefficient, C_L is the source concentration of the contaminant, and Q_L is the source discharge per unit length.

Equation (9.2.3) is a parabolic equation of the same form as (3.8.1) with extra terms.

Smith (1989) gives a useful discussion of the transitional regime where the rate of dilution of a contaminant increases from values associated with small-scale mixing to values associated with large-scale dispersion. Values of the diffusion coefficients given by Smith are:

(a) molecular diffusion 10^{-9} square metres per second;

(b) turbulent eddy diffusion 10^{-3} metres per second; and

(c) the longitudinal shear dispersion can be as large as 10^{3} metres per second.

Thus the significant diffusivity depends on the time or on the length scale. The molecular diffusion is usually negligible. For times longer than the cross-sectional mixing time, (c) is the appropriate value and the transport equation can be taken as one dimensional. Smith concludes that after several tidal cycles or over 100 channel breadths downstream of a discharge the transport can be taken as dominated by the longitudinal dispersion coefficient (c).

The following references are grouped according to the methods used to deal with the advection–diffusion equation as listed in section 9.1.

There is no example for the monolithic Method 1.

Method 2: single solute

Lui and Falconer (1989) solve the two-dimensional flow equations (1.4.5) and (1.4.8) with added terms to represent the bed shear stress and direct and lateral shear stress components, by the Falconer method (section 4.5). The pollutant transport equation for a single solute is solved by the QUICK scheme (section 3.8). They conclude from experiments that the QUICK scheme compares favourably with other methods particularly for modelling relatively high concentration gradients.

Method 2: multiple solutes

(a) Bach *et al.* (1989) use the one-dimensional St Venant equations
    (1.4.9) and (1.4.11) together with (9.2.3) with the source and decay
    terms to model a river in Denmark. The St Venant equations are
    solved by the Abbott scheme (section 4.5). The transport equation
    is solved by the QUICK scheme (section 3.8) for each of the
    dissolved pollutants. There are also the five differential equations
    giving the interactions of (1) oxygen, (2) BOD (Biological Oxygen
    Demand depending on the organic matter present), (3) ammonia,
    (4) nitrates, (5) temperature. These equations are solved separately
    at each time step using a fourth-order Runge–Kutta method
    (section 3.3). The model is calibrated using site data.

(b) Hooper *et al.* (1989) use an adaptation of the Falconer method
    (section 4.5) to model the River Exe estuary and the surrounding
    coastal region. This model also includes provision for wetting and
    drying in low water channels. The advection–diffusion is solved
    using the QUICKEST extension of the Leonard scheme (section
    3.8) to two dimensions (Ekebjaerg and Justesen, 1991). The water
    quality model also includes salinity effects. Vested *et al.* (1992)
    extend the QUICKEST scheme to three dimensions.

Method 3: single solute

(a) Adey and Brebbia (1973) give an early example of the Euler–
    Lagrange solution of the transport equation in two dimensions, in
    this case for the study of effluent dispersion in the Solent. This
    model uses observed velocities corresponding to the spring tide and
    diffusion coefficients based on experiments; thus there is only the
    transport equation (9.2.1) to be solved with source and decay terms
    added. The method of solution of the equation (9.2.1) is by using
    six-node finite elements and Euler– Lagrange operator splitting as
    follows.

    First the convection term is ignored and the equation is solved
    for one time step Δt as a diffusion equation using Crank–Nicolson
    (section 3.2). This is the Eulerian effect centred at a fixed point on
    the grid.

    Then the convection effect is calculated assuming that there is a
    fixed quantity of pollutant in each element and this is transported
    en bloc by the velocity components. This is the Lagrangian follow-
    ing of the particles.

    The new values of the concentration are then known at new
    positions and the values at the original nodal positions have to be

found by interpolation. The procedure is now repeated for the next time step.

(b) Cheng *et al.* (1984) also uses the Euler–Lagrange approach with the velocity components calculated by using ADI on a staggered grid (compare the predictor step of Falconer's method, section 4.6). From the knowledge of the velocity components the position of a point P is found such that with these velocity components a particle at P at time $n\Delta t$ would arrive at a grid point Q, $(i\Delta x, j\Delta y)$, at time $(n + 1)\Delta t$, i.e. P is the point

$$(i\Delta x - u_{ij}^{n+1}\Delta t, j\Delta y - v_{ij}^{n+1}\Delta t).$$

Then if the s, n are the coordinates along and perpendicular to the stream line PQ (Fig. 9.2.1) and a_L and a_T are the corresponding longitudinal and transverse dispersion parameters, the dispersion effect is calculated from

$$\frac{1}{\Delta t}[C_{ij}^{n+1} - C^n(\text{P})] = \frac{a_L}{\Delta s^2}[C^n(S_1) - 2C^n(\text{P}) + C^n(S_2)]$$

$$+ \frac{a_T}{\Delta n^2}[C^n(N_1) - 2C^n(\text{P}) + C^n(N_2)] \quad (9.2.4)$$

Cheng *et al.* simplify this by taking

$$\frac{\Delta t}{\Delta s^2}a_L = \frac{\Delta t}{\Delta n^2}a_T = 1$$

where Δs and Δn are called the dispersive spreading lengths of the concentration in the longitudinal and transverse directions in the time step Δt.

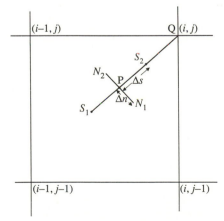

Figure 9.2.1: The method of Cheng *et al.* (1984).

Using the nine-point basis functions as given in section 6.3 the values of C at the points P, S_1, S_2, N_1, N_2 can be expressed in terms of the values at the grid points. Cheng *et al.* claim that this reduces numerical dispersion but there may be spurious oscillations.

(c) Sobey (1984) neglects the convective acceleration term and reduces the one-dimensional St Venant equations to

$$b \frac{\partial h}{\partial t} + \frac{\partial q}{\partial x} = 0 \qquad (9.2.5)$$

and

$$\frac{\partial q}{\partial t} + gA \frac{\partial h}{\partial x} = 0 \qquad (9.2.6)$$

which combine to form the wave equation

$$\frac{\partial^2 h}{\partial t^2} = \frac{gA}{b} \frac{\partial^2 h}{\partial x^2}. \qquad (9.2.7)$$

This form of the St Venant equations is used together with eqn (9.2.3) representing the advection–diffusion of a well-mixed waste with chemical reaction. Sobey discusses using the method of characteristics for the advection terms and the effect of different ways of interpolating back on to a fixed grid. He advocates the use of boxes displaced according to the advection with the diffusion etc. modelled by exchange between the boxes.

Method 3: multiple solutes

Osment *et al.* (1991) discuss numerical models to be used in the management of marine ecosystems. Irregular boundaries are treated by using body-fitting coordinates as in Reeve and Hiley (1992) with a transformation to a regular grid. The equations are then integrated over subregions and represented by finite difference expressions based on this grid. This model uses a central implicit finite difference scheme with the ADI method (section 4.5) to solve the hydrodynamic equations. The advection–diffusion equation is solved by an Euler–Lagrange-type method with the use of boxes as in Sobey (1984). The solute interaction equations are solved by the Bulirsch–Stoer extrapolation method (section 3.3). This method is very like the finite volume or cell-differencing method (section 3.11). In this, space is divided into cells with key quantities taken as cell averages; this means that conserved quantities are retained and fluxes are averaged over cell boundaries. Again it is necessary to point out that care is needed because non-uniform meshes may lead to loss of accuracy.

9.3 Subsurface flow—gradient recovery methods

In subsurface flow the velocity components must be obtained from the gradient of the solution for the piezometric head ϕ using the Darcy equations; in tensor form these are

$$v_i = -K_{ij}\frac{\partial\phi}{\partial x_j} \qquad (9.3.1)$$

$i, j = 1, 2$ for two dimensions, $i, j = 1, 2, 3$ for three dimensions (with summation over repeated suffixes). With finite difference methods the gradients can be obtained by differencing. With finite element methods there are differing opinions as to the best way to recover the gradients.

Yeh (1981) says that sampling the gradients at the nodes is unsatisfactory because of the different values in adjoining elements. He advocates solving the equations (9.3.1) separately with ϕ and the velocity components v_i expressed in terms of the same basis functions N_k:

$$\phi = \sum \phi_k N_k, \qquad v_i = \sum v_{ik} N_k. \qquad (9.3.2)$$

Mariño (1981) also advocates this global recovery method.

Nawalany (1990) uses a locally velocity-oriented approach. He solves for the piezometric head ϕ globally in a three-dimensional region and from this calculates the boundary conditions for particular subregions in which to solve eqns (9.3.1) for the velocity components.

The rationale for the global recovery methods is that in general the gradient is one order less accurate than the function; for example, with linear triangles the function has $O(h^2)$ error and the gradient only $O(h)$. With the global recovery using the same basis functions, the gradient is recovered to the same order of accuracy as the function.

Hawken *et al.* (1991) compare the global recovery method with a direct method in which gradients are recovered at particular points where they have been shown to be superconvergent as listed in section 7.5. They conclude from their experiments that the global recovery method and the use of superconvergent points give about the same accuracy but the use of the superconvergence is more cost-effective.

9.4 Water-quality modelling in subsurface flow

Mangold and Tsang (1991) give a comprehensive survey of 56 subsurface hydrological and hydrochemical numerical models from the United States and Canada. A count of the methods they list for solute transport models gives:

Saturated media, two dimensions
FEM (5), FDM (3), IFDM (1) FEM plus FDM (1);

Saturated media, three dimensions
FEM (1), FDM (3), IFDM (1);

Unsaturated media, two dimensions
FEM (3), FDM (0), IFDM (1), FEM plus IFDM (1);

Unsaturated media, three dimensions
FEM (1), FDM (2), IFDM (1), FEM plus IFDM (1)

where FEM denotes the finite element method (Chapter 6), FDM the finite difference method (Chapter 3), and IFDM the integrated finite difference method (section 5.4). This paper should be consulted for many other details.

In the proceedings of a conference on computational methods in subsurface hydrology in Venice 1990, the section on groundwater transport contamination problems has 11 out of 15 papers using FEM; the others are more concerned with management than with specific modelling methods.

In connection with water-quality modelling, where finite elements are used and the advection and diffusion terms are treated together, an upstream weighting technique (Huyakorn and Nilkuha, 1979; Huyakorn *et al.*, 1985) may be necessary. Here unsymmetric weighting functions are applied to the terms with space derivatives in (9.1.1) and the standard symmetric weights to the other terms. Kaluarachchi and Parker (1988) discuss this with reference to the transformation and transport of nitrogen species in the unsaturated subsurface zone.

Details of other methods mentioned in the literature are listed in the same categories as in section 9.1.

There are again no examples for Method 1.

Method 2: single solute

(a) Nishi *et al.* (1976) solve for the movement of a pollutant in seepage from a triangular channel (two dimensions in elevation). Eight-node isoparametric elements are used and there is no mention of upwinding.

(b) De Smedt (1990) uses isoparametric hexahedral elements for modelling subsurface transport of heavy metals for sludge disposal. There is no mention of upwinding, the mass matrix is lumped and the transport equation is solved by a predictor–corrector method.

(c) Pini *et al.* (1989) use upwinding with a three-dimensional problem solved by using tetrahedral elements. This paper is referred to again in Appendix 2 in connection with the preconditioned conjugate gradient method.

(d) Neuman (1984) uses an Euler–Lagrange method with a single solute. There is a distinction between the way particles are treated near steep concentration fronts and away from these. Neuman claims his method works with Peclet numbers up to infinity.

(e) Refsgaard and Jørgensen use the SHE model (Abbott *et al.*, 1980) to obtain the velocities for three-dimensional solute transport.

Rubin (1983), van Genuchten and Jury (1987), and Johnsson *et al.* (1987) discuss various aspects of the complications of subsurface water-quality modelling. Sposito *et al.* (1986) give some observations on the use of stochastic models. Hutson and Wagenet (1991) stress the importance of model sensitivity tests and methods for estimating data values with only a limited number of measurements.

9.5 Salt water intrusion problems

Coastal aquifers require careful management because of the danger of salt water intrusion. Methods for dealing with this problem range from simple to sophisticated. An example of the use of a simple numerical model for groundwater resource management is given by Spink and Wilson (1989) for an aquifer in South Humberside. The model is used to study the effects of abstraction wells and a reservoir. The aquifer is represented by a vertically integrated two dimensions in plan model as in eqn (1.7.10). The equation is solved by the finite difference method (Chapter 3). The boundary conditions are all no flow except for the seaward boundary. This is represented by fixed heads during the solution for the initial conditions and thereafter becomes no flow except for part of this boundary where there is a region of high permeability. The transmissivities based on the current head are calculated at each time step using eqns (1.7.8b) to allow for the variation of hydraulic conductivity with depth; there are three time steps per month. From the numerical solution for the piezometric head the flow on the high-permeability part of the seaward boundary is calculated. When this flow is inwards there will be a danger of saline intrusion. The model is used to predict the effects of redistributing abstraction wells and the use of a reservoir to support the flow during the summer months. The aim is to manage the aquifer so as to minimize the inflow from the sea.

The method just described is based on integrating through the vertical. More detail of the position of the salt water can be found by looking at the picture in two dimensions in elevation using the Ghyben–Herzberg approximation. Between the fresh water and sea water there is actually a transition zone in which the density varies from that of sea water to that of fresh water. The Ghyben–Herzberg approximation assumes instead that there is an abrupt interface between the salt and fresh water and that there is a dynamic steady state with a horizontal flow of fresh water over a stationary wedge of salt water (Fig. 9.5.1, not to scale). With horizontal flow $\partial\phi/\partial z = 0$, i.e. the equipotentials, $\phi = $ constant, are vertical. Then if h_f is the height of the phreatic surface above the sea level, h_s is the depth of the salt/fresh interface below this level, and ρ_s, ρ_f are the densities of the salt and fresh water respectively,

$$h_s\rho_s = (h_s + h_f)\rho_f$$

i.e.

$$h_s = \frac{\rho_f}{\rho_s - \rho_f} h_f = \alpha h_f \tag{9.5.1}$$

say. Thus with $\rho_s = 1.025$ g cm^{-3} and $\rho_f = 1.0$ g cm^{-3},

$$h_s = 40 h_f \tag{9.5.2}$$

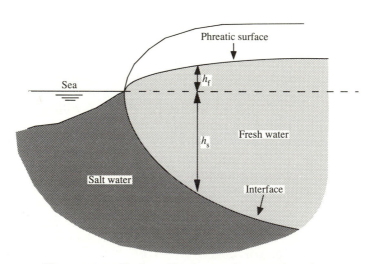

Figure 9.5.1: Ghyben–Herzberg salt water interface.

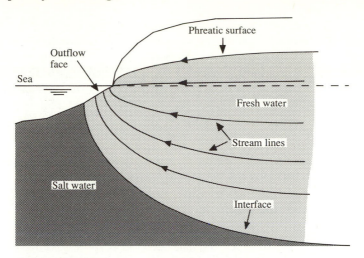

Figure 9.5.2: Salt water interface—stream lines near coast.

i.e. the depth of the stationary interface below sea level is 40 times the height of the phreatic surface above it. This approximation can be combined with the solution of the two dimensions in plan equation (1.7.10). During periods when the phreatic surface is only changing slowly the position of the salt water wedge can be calculated. 'The Ghyben–Herzberg approximation often comes surprisingly close to predicting the actual seawater interface' comments Frind (1980). Since the actual flow near the sea must have vertical components, it is better approximated as indicated in Fig. 9.5.2 (again not to scale). The actual depth of the interface is slightly larger than that given by the Ghyben–Herzberg assumption which is thus on the safe side. Bear (1979) gives a further discussion of this and the problem of upconing when an abstraction well is pumping from too near the interface.

More sophisticated numerical models take account of the movement of the salt water wedge and the transition zone between the salt and fresh water. If C is the dimensionless concentration of salt (in parts per thousand, for example) the transport equation is similar to (9.1.1):

$$\frac{\partial}{\partial t}(nC) + \nabla \cdot (n\mathbf{v}C) - \nabla(nD \cdot \nabla C) = 0 \qquad (9.5.3)$$

assuming no sources or sinks. The Darcy velocity equation in tensor form is now taken as

$$v_i = -\frac{k_{ij}}{\mu n}\left[\frac{\partial p}{\partial x_j} + \rho g e_j\right] \qquad (9.5.4)$$

where v_i is the ith component of the velocity, LT^{-1}, k_{ij} is the permeability tensor L^2, n is the porosity, x_j is the coordinate in the jth direction, μ is the dynamic viscosity, $ML^{-1}T^{-1}$, ρ is the density, ML^{-3}, ge_j is the (vertical) gravity effect, LT^{-2}, and p is the pressure $ML^{-1}T^{-2}$. It is simplest to take the hydraulic conductivity as given by

$$K_{ij} = \frac{k_{ij}}{\mu}\rho g \qquad (9.5.5)$$

with the values of μ and ρ for fresh water and use the freshwater head as the reference rather than use $\rho = \rho(C)$ and $\mu = \mu(C)$ in this expression. The form of the continuity equation also varies. It can depend on how you interpret the discussion in Bear (1979, pp. 92–93). Some people simply take

$$\nabla \cdot (\rho n \mathbf{v}) = 0 \qquad (9.5.6)$$

(Segol *et al.*, 1975; Wikramaratna and Wood, 1983), with $\rho = \rho_0(1 + \varepsilon C/C_1)$, where C_1 is the maximum concentration of the salt water, $\varepsilon = (\rho_1 - \rho_0)/\rho_0$ and ρ_1 is the maximum density of the salt water, ρ_0 is the density of the fresh water. Others take

$$\frac{\partial}{\partial t}(\rho n) + \nabla \cdot (\rho n \mathbf{v}) = 0 \qquad (9.5.7)$$

(Porter and Jackson, 1990). Another form is

$$\rho S_0 \frac{\partial \phi}{\partial t} = \nabla \cdot (\rho n \mathbf{v}) \qquad (9.5.8)$$

where S_0 is the specific storativity and the piezometric head is given by

$$\phi = \frac{1}{\rho_0 g} p + z, \quad (z = x_3) \qquad (9.5.9)$$

(Taylor and Huyakorn, 1976).

Whichever form of the continuity equation is chosen the net effect is a system of non-linear coupled equations for the concentration, the piezometric head, and the velocity components. The boundary conditions for the concentration are either known concentration or known concentration gradient. With a time-dependent problem the initial conditions must also be specified and as usual must be taken as near a steady state as possible to avoid introducing oscillation.

Henry (1959, 1964) solved the steady-state problem in two dimensions in elevation using a dimensionless stream function ψ' such that $u' = \partial\psi'/\partial y'$, $v' = \partial\psi'/\partial x'$ are dimensionless velocity components with dimensionless coordinates x', y'. Henry solved for ψ' and the concentration C in a rectangular region using Fourier series expansions. Huyakorn and Taylor (1976) made some numerical experiments to compare solutions using:

(1) u, v, pressure p, and concentration C;

(2) the stream function and concentration;

(3) the freshwater hydraulic head and concentration.

They concluded that the third method gives the fastest convergence in the iteration necessary to deal with the non-linearity. The stream function formulation is not discussed further here as it is not much used. The use of the freshwater head as listed in method 3 means that μ and ρ in (9.5.4) do not vary. Huyakorn and Taylor observed local oscillations where the Peclet number exceeded a critical value (for one-dimensional problems this critical number is 2). They obtained a smoother result by refining the mesh in a subregion using boundary values from the coarser mesh.

Further examples from the literature are the following.

Method 1: the monolithic approach

Porter and Jackson (1990) solve the equations for a two-dimensional problem *en bloc* with the emphasis on experiments with quasi-Newton methods (Appendix 2). They use quadrilateral elements and quadratic basis functions (it is not clear whether these are eight node or nine node).

Method 2

This class of methods has the velocities found from the flow equations and then the transport equation solved as an entity usually with some provision for upstream differencing and/or mesh refinement dependent on the mesh Peclet number.

Wikramaratna and Wood (1983) is an illustration of this type of method. This uses 'streamline upwinding' (Hughes and Brooks, 1979) to control the oscillations in a steady-state two dimensions in elevation salt water intrusion problem. The equations are solved for the pressure p and the salt concentration C using isoparametric bilinear elements. Figure 9.5.3 is a sketch of the region. The aquifer is unconfined and the position of the water table is initially unknown but the level is assumed known at an inland position where the groundwater is entirely fresh. With the nodes arranged in vertical lines and an

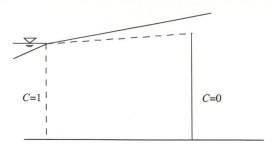

Figure 9.5.3: Sketch of region for salt water intrusion problem.

estimated position for the water table, the equations are first solved
with a no-flow condition on the top boundary and then for nodes j on
the free surface the height is adjusted by

$$z_{k+1}(j) = z_k(j) - \frac{p_k(j)}{g\rho_k(j)} \qquad (9.5.10)$$

where the subscript denotes the iteration number and the iteration is
repeated until the other boundary condition, $p = 0$, is satisfied. The
equations for the concentration are solved by iteration using a
relaxation parameter ω:

$$C_{k+1} = \omega C^*_{k+1} + (1 - \omega)C_k. \qquad (9.5.11)$$

Figure 9.5.4 shows the different effects in the convergence of using
$\omega = 1.0$ and 0.5. The averaging (underrelaxation) effect of using $\omega = 0.5$
gives much faster convergence here. Data used in the numerical
experiments are from the Mogadishu aquifer in Somalia (Wikramaratna,

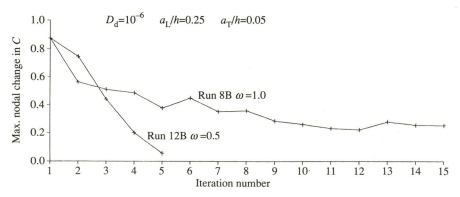

Figure 9.5.4: Convergence of iteration.

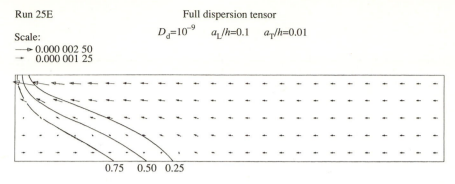

Run 25E   Full dispersion tensor

$D_\mathrm{d}{=}10^{-9}$ $a_\mathrm{L}/h{=}0.1$ $a_\mathrm{T}/h{=}0.01$

Scale:
→ 0.000 002 50
→ 0.000 001 25

0.75 0.50 0.25

Figure 9.5.5: Velocity vectors and isochlors for converged solution.

1981). Figure 9.5.5 (from Wikramaratna and Wood, 1983) shows the velocity vectors and isochlors $C = 0.25, 0.5, 0.75$. This is a typical flow pattern for salt water intrusion in a coastal aquifer with the salt water wedge effect represented by the recirculation zone in the bottom left-hand corner. The maximum velocity is at the top left-hand corner where the fresh water is flowing out over the wedge and is approximately horizontal here. Since the diffusion effect is dominated by the longitudinal dispersivity in the diffusion tensor, eqn (9.1.2), a convenient approximation to the grid Peclet number in this region of maximum velocity is given by

$$\frac{|v|h}{a_\mathrm{L}|v|} = \frac{h}{a_\mathrm{L}} \tag{9.5.12}$$

where h is the mesh size along the stream line. The numerical results show that this is the key parameter determining whether the solution for the concentration exhibits spatial oscillations or not. The steepest gradient in the salt concentration is also at the top left-hand corner of the region and this is used as a test; it should be monotonic. Table 9.5.1 (after Wikramaratna and Wood, 1983) lists results with various values of $a_\mathrm{L}, a_\mathrm{T}$, and the mesh size h. Figure 9.5.6 (from Wikramaratna, 1981) shows a typical mesh. As predicted the spatial oscillations occur when the parameter $h/(2a_\mathrm{L})$ is greater than unity. Making this parameter less than unity either by decreasing h or by increasing a_L removes the oscillations. Figure 9.5.7 (from Wikramaratna and Wood, 1983) shows the salt concentration variation near the corner where the salt gradient is a maximum, for (a) run 2/5 with the coarse mesh C and $h/(2a_\mathrm{L}) > 1$ showing oscillation, (b) run 2/6, and (c) run 2/4 with the oscillation removed by increasing a_L and decreasing h respectively. Figure 9.5.7 also shows the vertical variation of salt

Table 9.5.1

Mesh	Run no.	a_L	a_T	$h/(2a_L)$	Oscillation
A	2/3	10	1.0	0.417	No
A	2/4	5	0.5	0.833	No
A	2/10	4	0.5	1.042	Yes
B	2/2	10	1.0	0.833	No
B	2/6	10	0.5	0.833	No
B	2/5	5	0.5	1.667	Yes
C	2/9	15	0.5	0.833	No
C	2/8	10	0.5	1.25	Yes
C	2/7	5	0.5	2.5	Yes

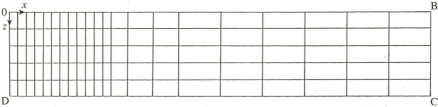

Figure 9.5.6: Typical mesh for salt water intrusion problem.

concentration at dimensionless distances of $x = 0.5$, 1.0 (corresponding to 50 m and 100 m inland with the Mogadishu data). An extension of this work to a three-dimensional coastal aquifer with abstraction wells using eight-node brick elements (Wikramaratna and Reeve, 1986), found that $\omega = 0.2$ was needed in (9.5.11) for fast convergence when the abstraction wells were operating.

Method 3: Euler–Lagrange split of the transport equation

Gambolati *et al.* (1990) use the Euler–Lagrange method of Neuman (1984) in solving a salt water intrusion problem. They remark that the dependence of the flow velocity on the concentration C is weak and they compare the results of iteration to convergence with those obtained with only one iteration for each equation. They say the latter does not introduce appreciable errors in the values of the concentration.

See also in connection with water quality modelling the very general discussion in the ASCE Task Committee (1988) series of papers on turbulence modelling in surface water flow and transport. The last of this series of five papers includes many useful references.

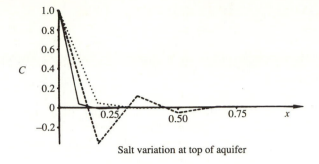

Salt variation at top of aquifer

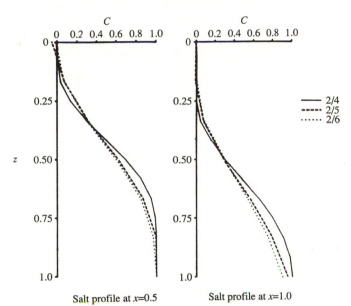

Salt profile at *x*=0.5   Salt profile at *x*=1.0

Figure 9.5.7: Effect of $h/(2a_{\mathrm{L}})$ on salt profiles.

10   Dimensionless parameters

10.1 Introduction

So far in this book the equations have mainly been left in their dimensional form but in practice the equations should always be non-dimensionalized before the numerical method is applied. This gives a clue as to the relative sizes of the terms in the equations and hence guidance as to the strategy to be used. For example, taking a more general approach than that in section 1.4, the Navier–Stokes equation for one-dimensional surface flow assuming an incompressible fluid with laminar flow and constant density and viscosity is (see e.g. French, 1987)

$$\frac{\partial u}{\partial t} + u\frac{\partial u}{\partial x} + g\frac{\partial h}{\partial x} + \frac{1}{\rho}\frac{\partial p}{\partial x} = \frac{\mu}{\rho}\frac{\partial^2 u}{\partial x^2} \qquad (10.1.1)$$

where p is the pressure and μ the viscosity and otherwise the notation is the same as previously. Equation (10.1.1) can be non-dimensionalized by substituting

$$x = \hat{x}L, \quad u = \hat{u}U, \quad t = \hat{t}L/U, \quad p = \hat{p}\rho U^2, \quad h = \hat{h}L \qquad (10.1.2)$$

where L, U are a characteristic length and velocity. Substituting from (10.1.2) and multiplying through by L/U^2 gives the non-dimensionalized equation

$$\frac{\partial \hat{u}}{\partial t} + \hat{u}\frac{\partial \hat{u}}{\partial \hat{x}} + \frac{gL}{U^2}\frac{\partial \hat{h}}{\partial \hat{x}} + \frac{\partial \hat{p}}{\partial \hat{x}} = \frac{\mu}{\rho LU}\frac{\partial^2 \hat{u}}{\partial \hat{x}^2}. \qquad (10.1.3)$$

Substituting the Froude number $\mathbf{F} = U/\sqrt{(gL)}$ and the Reynolds number $\mathbf{R}_e = \rho UL/\mu = UL/v$, where v is the kinematic viscosity, then gives

$$\frac{\partial \hat{u}}{\partial t} + \hat{u}\frac{\partial \hat{u}}{\partial \hat{x}} + \frac{1}{\mathbf{F}^2}\frac{\partial \hat{h}}{\partial \hat{x}} + \frac{\partial \hat{p}}{\partial \hat{x}} = \frac{1}{\mathbf{R}_e}\frac{\partial^2 \hat{u}}{\partial \hat{x}^2}. \qquad (10.1.4)$$

In surface flow the Reynolds number is the basis for the classification of the flow as laminar, transitional, or turbulent (see e.g. French, 1987). In open channel flow the characteristic length usually used in this classification is the hydraulic radius, i.e. the ratio of the cross-sectional area of the flow to the wetted perimeter (section 1.2). The characteristic velocity U is taken as the average velocity of the flow. A very large value of $\mathbf{R}_e$ indicates that the second-derivative term in (10.1.4) can be neglected and the equation becomes hyperbolic.

The Froude number $\mathbf{F}$ is the ratio of inertial to gravity forces and the flow is classed as subcritical, critical, or supercritical according to whether $\mathbf{F}$ is less than, equal to, or greater than unity. French (1987) gives the characteristic length here as the hydraulic depth which is equal to the flow area divided by the width of the free surface, so not precisely the same as that used in the Reynolds number.

Woolhiser and Liggett (1967) described experiments to determine when it is sufficient to solve the kinematic equation rather than the full St Venant equations. Their results depend on a dimensionless parameter k (sometimes called the kinematic wavenumber):

$$k = \frac{sL}{h_L \mathbf{F}_L^2} \tag{10.1.5}$$

where s is the slope, L is the length of the slope, h_L is the depth at $x = L$ when the discharge is at a maximum and the velocity is u_L, and $\mathbf{F}_L = u_L/\sqrt{(gh_L)}$ is the corresponding Froude number. Their conclusion is that the kinematic wave equation is good enough when $k > 10$. Morris and Woolhiser (1980) add on to this the condition that for very low values of $\mathbf{F}_L$ a condition which also needs to be satisfied is

$$\mathbf{F}_L^2 k = \frac{sL}{h_L} \geq 5. \tag{10.1.6}$$

Limitations on the use of the box scheme discussed by Samuels and Skeels (1990) (section 4.4) relate to the size of the Vedernikov number

$$\mathbf{V} = \frac{m\mathbf{F}A}{nR}\frac{dR}{dA} \tag{10.1.7}$$

with the notation as in section 1.2 and $\mathbf{F}$ is the Froude number. French (1987) defines the Vedernikov number as

$$\mathbf{V} = \frac{m\mathbf{F}}{n}\left(1 - R\frac{dP}{dA}\right). \tag{10.1.8}$$

Since the cross-sectional area, A, the hydraulic radius R, and the wetted perimeter P are related by

$$PR = A$$

differentiating gives

$$P\frac{dR}{dA} + R\frac{dP}{dA} = 1$$

i.e.

$$1 - R\frac{\mathrm{d}P}{\mathrm{d}A} = P\frac{\mathrm{d}R}{\mathrm{d}A} = \frac{A}{R}\frac{\mathrm{d}R}{\mathrm{d}A}.$$

Hence (10.1.7) and (10.1.8) are equivalent.

In subsurface flow it is only the continuity equation which is used together with the substitution for the specific discharge from Darcy's law. Bear (1979) states that Darcy's law is valid so long as the Reynolds number $\mathbf{R}_\mathrm{e}$ does not exceed some value between 1 and 10. $\mathbf{R}_\mathrm{e}$ is defined here as

$$\mathbf{R}_\mathrm{e} = qd/v \qquad\qquad (10.1.9)$$

where q is the magnitude of the specific discharge, d is some representative length of the porous matrix (usually a representative grain size), and v is the kinematic viscosity of the fluid. For this range of $\mathbf{R}_\mathrm{e}$ the viscous forces are predominant and the flow is laminar; most groundwater flows are within this range.

The mass-transport equations for surface and subsurface flows are represented by eqn (9.1.1) where the parameter n is the porosity for groundwater flow and unity for surface flow. The one-dimensional transport equation is

$$\frac{\partial}{\partial t}(AC) + \frac{\partial}{\partial x}(uAC) - \frac{\partial}{\partial x}\left(AD\frac{\partial C}{\partial x}\right) = -FAC + Q_L C_L. \quad (10.1.10)$$

To non-dimensionalize put

$$x = \hat{x}L, \quad A = \hat{A}L^2, \quad u = \hat{u}\bar{u}, \quad t = \hat{t}L/\bar{u}, \quad D = \bar{D}\hat{D} \quad (10.1.11)$$

where $\bar{u}, \bar{D}$ are average values of the velocity and coefficient of diffusion. The result after dividing through by $L\bar{u}$ is

$$\frac{\partial}{\partial \hat{t}}(\hat{A}C) + \frac{\partial}{\partial \hat{x}}(\hat{u}\hat{A}C) - \frac{\bar{D}}{L\bar{u}}\frac{\partial}{\partial \hat{x}}\left(\hat{A}\hat{D}\frac{\partial C}{\partial \hat{x}}\right) = -\frac{F\hat{A}LC}{\bar{u}} + \frac{Q_L C_L}{L\bar{u}}. \quad (10.1.12)$$

The dimensionless parameter $L\bar{u}/\bar{D}$ represents the ratio of advection to diffusion. In connection with subsurface flow Bear (1979) calls $L\bar{u}/\bar{D}$ the Peclet number with L here a characteristic length of the pores and $\bar{D}$ the coefficient of molecular diffusion. In general the molecular diffusion effect is absorbed into the mechanical diffusion/dispersion.

In the numerical solution the characteristic length is the length of the channel in the flow represented in eqn (10.1.10) or the width of the channel in the corresponding non-dimensionalization of the problem of lateral velocity distribution in section 1.3. In multi-dimensional problems the characteristic length is usually the largest dimension.

The result must be that Δx, Δy, Δz, or h, the finite element size, are fractions of the dimensions of the region. The time may be non-dimensionalized with respect to a characteristic time rather than with a relation involving length and velocity as in (10.1.2). The characteristic time depends on the time span of the events being studied, which may be hours, days, or years. The time step Δt must be a fraction of the time span.

In the numerical methods described in this book the following dimensionless parameters have appeared: Courant number $C_N = c\Delta t/\Delta x$, section 3.10, in connection with the CFL condition and conditional stability, $C_N \leq 1$; mesh Peclet number $P_e = u\Delta x/D$ (section 3.8) in connection with the need for upstream differencing; $P_e \leq 2$ is the condition to avoid spurious oscillations with the central difference scheme.

In the next section a more general approach to the formation of dimensionless parameters is described.

10.2 The Buckingham pi theorem

This is a method for dealing with any problem where there is a confusing number of parameters. It is illustrated on the particular case of combined surface and subsurface flow due to rainfall on a hillslope taken from Calver and Wood (1991). The supposition here is that there is a unit-independent physical law giving the discharge per unit width Q at the bottom of the hillslope in terms of the time t, the duration of the rainfall t_r, the rate of rainfall r (the effective precipitation, after evaporation and interception), the saturated hydraulic conductivity K_s, the length L and the depth l of the subsurface region, the initial value ψ_{in} of the pressure potential ψ in the region, the slope s, the porosity θ_s and the surface roughness f. The parameters which determine the non-linear variation of K/K_s and θ/θ_s in the unsaturated zone are left fixed for this study. There is a free seepage face at the bottom of the slope.

The Buckingham pi theorem (Buckingham, 1914, 1915) states that this physical law is equivalent to an equation expressed only in dimensionless quantities. All the physical quantities just listed are either dimensionless (s and θ_s) or can be expressed in the basic dimensions of length L and time T. The dimensions of the surface roughness f are derived from the Chezy formula so that they are $L^{1/2}T^{-1}$; see the section on notation for the dimensions of the other quantities. There are thus nine quantities Q, t, t_r, r, K_s, L, l, ψ_{in}, and f which are functions of length and time. In general, for n quantities which are functions of m dimensions, the physical law can be expressed in terms of $(n - m)$ independent dimensionless quantities; hence in this problem there are seven. They can be chosen in many ways but they must be independent, i.e. there must be no functional relationship between them. Using the Buckingham pi notation, $\pi_1, \pi_2, \ldots$ for the dimensionless parameters it seems natural to start with $\pi_1 = t/t_r$ (to link the two times), $\pi_2 = L/l$ (to scale the subsurface region), and $\pi_3 = -\psi_{in}/l$ (the parameters must be positive and it seems natural to link ψ_{in} with the depth rather than the length). The other parameters are not so obvious but as the result of experiment the most convenient choice is taken as $\pi_4 = Q/rL$ (relating the discharge to the rate at which the rainfall arrives on the whole length of the slope), $\pi_5 = rt_r/L$ (the amount of rain per unit length of slope), $\pi_6 = r/K_s$ (Buckingham recommends taking ratios of like quantities), and $\pi_7 = f^2t_r/r$ (the surface roughness, having slightly awkward dimensions, is restricted to one parameter and this relates it to the duration and intensity of the rainfall).

The quantity Q which it is desired to express in terms of the other variables is present in one π parameter only, π_4; this and only one other parameter, π_1, are functions of time. By the Buckingham pi theorem the dimensionless equation can be expressed in the form

$$\pi_4 = Q/rL = G(\pi_1, \pi_2, \pi_3, \pi_5, \pi_6, \pi_7) \qquad (10.2.1)$$

where $G(\ldots)$ is some unspecified functional relationship. Differentiating (10.2.1) with respect to time gives

$$\frac{1}{rL} \dot{Q} = \frac{\partial G}{\partial \pi_1} \frac{d\pi_1}{dt} = \frac{1}{t_r} \frac{\partial G}{\partial \pi_1}. \qquad (10.2.2)$$

There is particular interest in the peak discharge Q_p and the time t_p at which this arrives. The peak discharge is when $\dot{Q} = 0$, i.e. when $\partial G/\partial \pi_1 = 0$ and $t = t_p$. This is equivalent to another dimensionless statement which can be written in the form

$$t_p/t_r = F(\pi_2, \pi_3, \pi_5, \pi_6, \pi_7) \qquad (10.2.3)$$

(another unspecified functional relationship). Equation (10.2.1) also gives

$$Q_p/rL = G(t_p/t_r, \pi_2, \pi_3, \pi_5, \pi_6, \pi_7). \qquad (10.2.4)$$

Hence substituting from (10.2.3) into (10.2.4) gives

$$Q_p/rL = H(\pi_2, \pi_3, \pi_5, \pi_6, \pi_7). \qquad (10.2.5)$$

This is also an unspecified functional relationship but (10.2.3) and

(10.2.5) can be used as the basis for numerical results to compare with site data. The numerical results are obtained using the IHDM model for surface and subsurface flow described in section 7.7. Each simulation is allowed to run in until the base flow Q_b is changing only very slowly before the rainfall starts so that any spurious fluctuation has disappeared and the rise $Q_p - Q_b$ is plotted.

In expressing results in terms of groupings of dimensionless parameters a single result may be achieved from a variety of combinations of the physical parameters. But when the modelling is done numerically it is important to check the effect of the discretization of the problem on the degree of scatter of the results. In the experiments described in Calver and Wood (1991) it is found necessary to use the most refined elements referred to as the baseline grid in section 7.8. In the results shown in Figs 10.2.1–4 (from Calver and Wood, 1991) the rate of rainfall

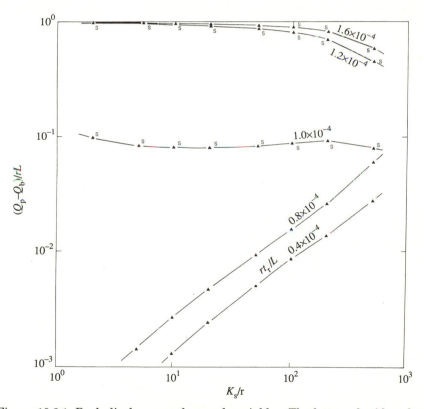

Figure 10.2.1: Peak discharge and causal variables. The letter s beside a data point indicates that the discharge includes saturation-excess overland flow in addition to subsurface flow.

r is kept constant at 5 mm per hour and the duration of rainfall t_r and the saturated hydraulic conductivity K_s are varied. Other parameters are $s = 0.1$, $L = 500$ m, $l = 1.5$ m, $\theta_s = 0.4$, and Chezy surface roughness 2×10^4 $m^{0.5}$ per hour. The relationships of hydraulic conductivity and water content to pressure potential in the unsaturated region are assumed to be single-valued functions representative of a medium-textured soil taken from Clapp and Hornberger (1978). Saturated throughflow discharge comes from the seepage face which is taken to be above channel water level. Two commonly considered initial conditions are used, the same as those detailed in section 7.9.

In Figs 10.2.1 and 10.2.2 ψ_{in} is everywhere set to -0.3 m which leads to a saturated layer all the way up the slope by the time the rainfall is introduced after the run-in period. In Figs 10.2.3 and 10.2.4 the initial pressure potential at a point is set at one-tenth of the height of that point above the foot of the slope which leads to a saturated wedge. Figures 10.2.1 and 10.2.3 show $(Q_p - Q_b)/(rL)$ (ratio of peak discharge rise above baseflow to the rainfall input, each per unit width) and Figs 10.2.2 and 10.2.4 show the ratio of time to peak to the duration of the

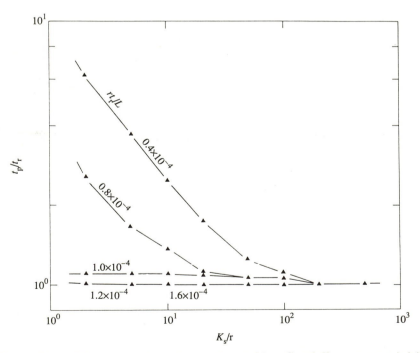

Figure 10.2.2: Time to peak and causal variables. Spatially constant initial conditions.

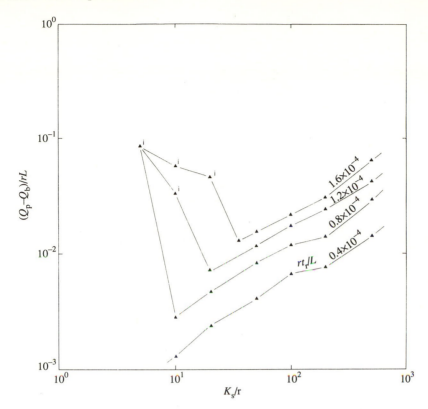

Figure 10.2.3: Peak discharge and causal variables. The letter i beside a data point indicates that runoff includes infiltration-excess overland flow in addition to subsurface flow.

rainfall. In each figure the results are plotted against K_s/r (the ratio of rate of subsurface travel to the rate of water addition) for various values of rt_r/L (ratio of total depth of rainfall to the slope length). The segments of the curves are drawn as straight lines between the calculated points.

In Fig. 10.2.1 all the discharge results involve saturated throughflow at the foot of the slope. The three uppermost curves also involve surface flow derived from oversaturation of the soil profile over part of the slope. When there is surface flow there are higher values of $(Q_p - Q_b)/(rL)$ which can approach unity for lower K_s/r. Figure 10.2.2 shows the corresponding times to peak ratios. The time to peak tends to the time at which the rain stops as K_s/r increases and this minimum time is reached for lower K_s/r for those cases where there is surface flow. Figures 10.2.3 and 10.2.4 show the corresponding results with the

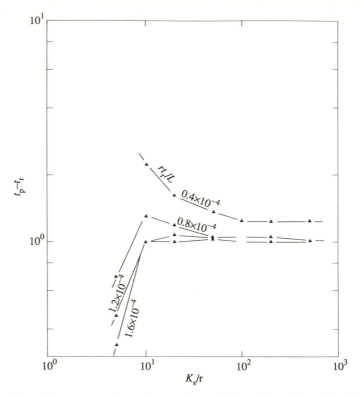

Figure 10.2.4: Time to peak and causal variables. Elevation-related initial
conditions.

initial wedge. In Fig. 10.2.3 there is the effect of infiltration-excess but
not saturation-excess surface flow. With the initial wedge the upper
part of the slope is drier and infiltration capacity is reduced; also there
are higher subsurface conductivities produced by the wetter conditions
near the lower part of the slope. Thus in Fig. 10.2.4 there are some t_p/t_r
values less than 1 for the low K_s/r values showing the effect of
infiltration-excess surface flow early in the storm event and the
domination of the peak discharge by this rather than the later
throughflow. Calver and Wood (1991) give details of two field examples
used to demonstrate the use of this dimensionless parameter approach.
It can be used:

(1) to give a quick estimation of runoff, either for a particular event
    or for a design storm;

(2) to determine an operative value of a parameter, for example the hydraulic conductivity, given measured values of the peak discharge and time to peak. Separate consideration of a number of events is needed to check the convergence on to a particular value.

Appendix 1　Vertical integration of groundwater flow equations

---

The three-dimensional continuity equation for groundwater flow is obtained in section 1.7 as

$$S_0 \frac{\partial \phi}{\partial t} = -\operatorname{div} \mathbf{q} = -\frac{\partial q_x}{\partial x} - \frac{\partial q_y}{\partial y} - \frac{\partial q_z}{\partial z} \qquad (A1.1)$$

with coordinate axes $0x, 0y, 0z$ with $0z$ vertically up. The physical situation considered here is that of a phreatic, i.e. unconfined, aquifer. The ground is saturated up to the water table, the phreatic surface. First the boundary conditions on the top and bottom of the aquifer are obtained and then eqn (A1.1) is integrated through the vertical using Leibniz's rule.

Boundary condition on the phreatic surface

A useful general rule given by Bear (1979) is that from a statement that a physical property

$$G = \text{constant} \qquad (A1.2)$$

must follow the equation

$$\frac{\mathrm{D}G}{\mathrm{D}t} \equiv \frac{\partial G}{\partial t} + \mathbf{v} \cdot \nabla G = 0 \qquad (A1.3)$$

where $\mathrm{D}/\mathrm{D}t$ is the total derivative and $\mathbf{v}$ is the velocity of the 'particles carrying the property G'.

If the phreatic surface is given by the equation $h(x, y, t) = z$, this can be written as

$$G = h(x, y, t) - z = 0. \qquad (A1.4)$$

Suppose the rate of accretion per unit area of the phreatic surface is represented by the vector $\mathbf{N}$. Then

$$S\mathbf{v} = \mathbf{q} - \mathbf{N} \qquad (A1.5)$$

where S is the effective storativity, $\mathbf{q}$ is the specific discharge of the fluid at a point P on the phreatic surface, and $\mathbf{v}$ is the velocity of propagation of the fluid here. Substituting in (A1.3) gives

$$S\frac{\partial h}{\partial t} + (\mathbf{q} - \mathbf{N})\cdot\nabla(h - z) = 0 \tag{A1.6}$$

i.e.

$$S\frac{\partial h}{\partial t} + q_x\frac{\partial h}{\partial x} + q_y\frac{\partial h}{\partial y} - q_z + N = 0 \tag{A1.7}$$

where N is the numerical value of the rate of accretion.

Boundary condition at the base of the aquifer

If $\hat{\mathbf{n}}_b$ is the unit normal to the base of the aquifer then the equation

$$\mathbf{q}\cdot\hat{\mathbf{n}}_b = 0 \tag{A1.8}$$

expresses the condition that there is no flow across this surface. If the base is given by the equation

$$G_b = h_b(x, y) - z = 0 \tag{A1.9}$$

then

$$\hat{\mathbf{n}}_b = \frac{\nabla G_b}{|\nabla G_b|} = \frac{1}{|\nabla G_b|}\left[\frac{\partial h_b}{\partial x}, \frac{\partial h_b}{\partial y}, -1\right]^{\mathrm{T}}. \tag{A1.10}$$

Hence substituting into (A1.8) gives the boundary condition on the base of the aquifer:

$$q_x\frac{\partial h_b}{\partial x} + q_y\frac{\partial h_b}{\partial y} - q_z = 0. \tag{A1.11}$$

Now eqn (A1.1) is integrated through the vertical from $z = h_b$ to $z = h$ to give

$$\int_{h_b}^{h} S_0\frac{\partial \phi}{\partial t}\,dz = -\int_{h_b}^{h}\left[\frac{\partial q_x}{\partial x} + \frac{\partial q_y}{\partial y} + \frac{\partial q_z}{\partial z}\right]dz$$

$$= -\int_{h_b}^{h}\left[\frac{\partial q_x}{\partial x} + \frac{\partial q_y}{\partial y}\right]dz - [q_z]_{h_b}^{h}. \tag{A1.12}$$

Leibniz's rule for differentiation under the integral sign is now used as follows:

$$\int_{h_b}^{h}\frac{\partial q_x}{\partial x}\,dz = \frac{\partial}{\partial x}\int_{h_b}^{h}q_x\,dz - q_x(x, y, h, t)\frac{\partial h}{\partial x} + q_x(x, y, h_b, t)\frac{\partial h_b}{\partial x} \tag{A1.13}$$

and similarly for

$$\int_{h_b}^{h} \frac{\partial q_y}{\partial y} \, dz.$$

Substituting in (A1.12) then gives

$$\int_{h_b}^{h} S_0 \frac{\partial \phi}{\partial t} \, dz = -\frac{\partial}{\partial x} \int_{h_b}^{h} q_x \, dz - \frac{\partial}{\partial y} \int_{h_b}^{h} q_y \, dz$$

$$+ q_x(x, y, h, t) \frac{\partial h}{\partial x} - q_x(x, y, h_b, t) \frac{\partial h_b}{\partial x}$$

$$+ q_y(x, y, h, t) \frac{\partial h}{\partial y} - q_y(x, y, h_b, t) \frac{\partial h_b}{\partial y}$$

$$- q_z(x, y, h, t) + q_z(x, y, h_b, t). \tag{A1.14}$$

The boundary conditions (A1.7) and (A1.11) can now be used to simplify (A1.14) into

$$S \frac{\partial h}{\partial t} + \int_{h_b}^{h} S_0 \frac{\partial \phi}{\partial t} \, dz = -\frac{\partial}{\partial x} \int_{h_b}^{h} q_x \, dz - \frac{\partial}{\partial y} \int_{h_b}^{h} q_y \, dz + N \tag{A1.15}$$

For most practical purposes $S \gg S_0(h - h_b)$. Hence the left-hand side of (A1.15) is approximated by $S \, \partial h/\partial t$ and using Darcy's law (1.7.1) in the right-hand side finally gives the vertically integrated flow equation as

$$S \frac{\partial h}{\partial t} = \frac{\partial}{\partial x} \int_{h_b}^{h} K_x \frac{\partial \phi}{\partial x} \, dz + \frac{\partial}{\partial y} \int_{h_b}^{h} K_y \frac{\partial \phi}{\partial y} \, dz + N \tag{A1.16}$$

where K_x, K_y are the components of the hydraulic conductivity.

The three-dimensional equation (A1.1) can be expressed as an equation for the hydraulic potential ϕ by using Darcy's law to substitute for q_x, q_y, q_z. It is complicated to solve this equation up to the phreatic surface whose position is initially unknown. The position of the surface has to be determined by the two boundary conditions (A1.7) and $\phi = z$. There is also the complication of the capillary fringe effect which needs to be included in some cases (see section 7.9). The three-dimensional equation can be solved up to the surface of the ground with some empirical relation to give the value of the hydraulic conductivity in the unsaturated zone. This is Richards' equation and the solution is discussed in section 7.7. This has to include the surface flow from the effective precipitation which does not enter the ground plus water emerging when the surface of the ground is saturated.

The two dimensions in plan-vertically-integrated equation (A1.15) includes the boundary conditions at the top and bottom of the aquifer. The next step is to consider how to approximate the integrals in this equation. Dupuit (1863) observed that the slope of the phreatic surface is generally small (of the order of 1/1000 or 1/100), and hence it is reasonable to assume that the flow in an unconfined aquifer with an approximately horizontal base is itself essentially horizontal. This assumption leads to the result that the components of the specific discharge can be taken as

$$q_x = -\bar{K}_x \frac{\partial h}{\partial x}, \qquad q_y = -\bar{K}_y \frac{\partial h}{\partial y} \qquad (A1.17)$$

where $h = h(x, y, t)$ and $\bar{K}_x$, $\bar{K}_y$ are the average values of the hydraulic conductivity components over the vertical. Then q_x, q_y are not functions of z and hence the integrals in (A1.15) can be replaced to give the non-linear equation

$$S \frac{\partial h}{\partial t} = \frac{\partial}{\partial x} \left(\bar{K}_x (h - h_b) \frac{\partial h}{\partial x} \right) + \frac{\partial}{\partial y} \left(\bar{K}_y (h - h_b) \frac{\partial h}{\partial y} \right) + N \quad (A1.18)$$

(see Bear (1979) for a more detailed discussion). If the base of the aquifer can be taken as horizontal this can be used as datum, i.e. $h_b = 0$. With the further assumption that $\bar{K}_x h = T_x$, $\bar{K}_y h = T_y$ where T_x, T_y are the transmissivities obtained from site pumping tests the equation has the form

$$S \frac{\partial h}{\partial t} = \frac{\partial}{\partial x} \left(T_x \frac{\partial h}{\partial x} \right) + \frac{\partial}{\partial y} \left(T_y \frac{\partial h}{\partial y} \right) + N. \qquad (A1.19)$$

The term N here is the vertical component of the surface rate of accretion from the phreatic surface boundary condition in (A1.7). If it is sufficient to consider only the steady state and the variation in h is sufficiently small to treat the transmissivities T_x, T_y as independent of h, the problem is represented by the linear elliptic equation

$$\frac{\partial}{\partial x} \left(T_x \frac{\partial h}{\partial x} \right) + \frac{\partial}{\partial y} \left(T_y \frac{\partial h}{\partial y} \right) = 0 \qquad (A1.20)$$

If there are recharge or abstraction wells in the aquifer whose dimensions are small compared with those of the aquifer being modelled, these are conveniently represented by Dirac delta generalized functions as described in section 5.2.

Equation (A1.19) is the form of the groundwater flow equation most suitable for solution by finite differences as described in Chapter 5. Examples of the solution of the other forms of the equation by finite elements are discussed in Chapter 7.

Appendix 2   Methods for solving simultaneous equations

This appendix contains (1) a list of methods for solving systems of linear equations with references, (2) some notes on solving non-linear equations with references.

1. Most of the methods for the solution of the differential equations described in this book lead to systems of equations of the form

$$\mathbf{Au} = \mathbf{b} \tag{A2.1}$$

where the matrix $\mathbf{A}$ is usually symmetric, banded and often sparse (i.e. has many zero entries) within the band. When this system is linear it may be solved by one of two basic methods: (a) direct, (b) iterative.

(a) The most used direct methods involve forward elimination and back substitution, e.g. Gauss, Cholesky (for symmetric matrices), or algorithms for tridiagonal or quindiagonal matrices (used with alternating direction methods). One disadvantage with direct methods used on matrices which are not so simple as the tri- or quindiagonal matrices is that they need some special node-numbering scheme and pivoting strategy to make the algorithm computationally efficient. Even then storage requirements can still be excessively large especially for three-dimensional problems. Another disadvantage can be that the many arithmetic operations can lead to the accumulation of round-off errors for some types of matrix (Conte and de Boor, 1980).

(b) Iterative schemes avoid the need for storage of large matrices. The standard iterative schemes include Jacobi, Gauss–Seidel, and successive over relaxation (SOR) (Conte and de Boor, 1980). These generate a sequence of approximate solutions $\mathbf{u}^{(k)}$ by means of equations of the form

$$\mathbf{Mu}^{(k+1)} = \mathbf{Nu}^{(k)} + \mathbf{b} \tag{A2.2}$$

where the matrices $\mathbf{M}, \mathbf{N}$ come from different splittings of the matrix $\mathbf{A}$ (Varga, 1962, p. 88). The efficiency of these methods depends on a good starting value and, with SOR, a good estimate of the optimum value for the SOR parameter, ω to speed up the convergence, but it is usually difficult to determine this value. The conjugate gradient method does not have these difficulties. Used with an Nth-order symmetric positive definite matrix $\mathbf{A}$ it theoretically makes the residual $\mathbf{r} = \mathbf{Au} - \mathbf{b}$ zero after N steps. But rounding errors can accumulate so that this is not guaranteed and anyway N may be very large. Hence the present emphasis on the preconditioned conjugate gradient (PCG) methods where the system (A2.1) is preconditioned by being multiplied through by a matrix $\mathbf{C}^{-1}$ which gives a system for which the method converges much faster. The result of premultiplying (A2.1) by $\mathbf{C}^{-1}$ is

$$\mathbf{C}^{-1}\mathbf{AC}^{-1}(\mathbf{Cu}) = \mathbf{C}^{-1}\mathbf{b}. \qquad (A2.3)$$

Hence if $\tilde{\mathbf{A}} = \mathbf{C}^{-1}\mathbf{AC}^{-1}$, $\mathbf{x} = \mathbf{Cu}$, $\tilde{\mathbf{b}} = \mathbf{C}^{-1}\mathbf{b}$, the system becomes

$$\tilde{\mathbf{A}}\mathbf{x} = \tilde{\mathbf{b}}. \qquad (A2.4)$$

$\tilde{\mathbf{C}}$ is a symmetric positive definite matrix chosen so that $\tilde{\mathbf{A}}$ is a matrix with eigenvalues clustered round unity. Some people simply divide through the equations in the original system (A2.1) so that the diagonal entries are all unity. A more effective but surprisingly simple way is to obtain an approximate Cholesky decomposition of $\mathbf{A}$:

$$\mathbf{A} \approx \mathbf{HH}^{\mathrm{T}} = \mathbf{M} \qquad (A2.5)$$

by going through the Cholesky algorithm putting $h_{ij} = 0$ if $a_{ij} = 0$. Then the algorithm is reformed so that it uses the matrix $\mathbf{M}$ and the operation is repeated until the residual is zero.

Kuiper (1987) compares the efficiency of 17 different iterative methods for the solution of the non-linear three-dimensional groundwater flow equations and concludes that the preconditioned conjugate gradient method is the best. See, for example, Pini *et al.* (1989).

See Conte and de Boor (1980) for the more elementary methods mentioned here; Duff *et al.* (1989) for direct methods; and Golub and van Loan (1990) for direct and iterative methods including a full discussion of PCG methods.

2. Non-linear equations such as that arising in the solution of Richards' equation in Chapter 7 can often be solved by some iteration method as in the Cooley algorithm described in section 7.7. Once a time-stepping scheme has been started, provided the time step Δt is reasonably small, there is a good starting value for the iteration available in either the solution from the last time step or some extrapolation such as in (7.7.8). For a steady-state problem it may not be so simple to get a good start to the iteration.

The method of Newton (Newton–Raphson) can give fast convergence to the solution of a system of nonlinear equations. These are written in the form

$$\mathbf{f}(\mathbf{u}) = 0. \tag{A2.6}$$

Then if $\hat{\mathbf{u}}$ is an approximate solution and the exact solution is $\hat{\mathbf{u}} + \delta\hat{\mathbf{u}}$, using an extension of Taylor's series gives

$$0 = \mathbf{f}(\hat{\mathbf{u}} + \delta\hat{\mathbf{u}}) = \mathbf{f}(\hat{\mathbf{u}}) + \mathbf{f}'(\hat{\mathbf{u}})\delta\hat{\mathbf{u}} + \theta\|\delta\hat{\mathbf{u}}\|^2 \tag{A2.7}$$

where $\delta\hat{\mathbf{u}}$ is a vector of increments in $\hat{\mathbf{u}}$, the matrix $\mathbf{f}'(\hat{\mathbf{u}})$, called the Jacobian matrix, has i, j entries given by

$$\{\partial f_i / \partial \hat{u}_j\} \tag{A2.8}$$

and the last term on the right of (A2.7) contains squares of the increments $\delta\hat{u}_i$. Neglecting these second-order terms gives the linearized equation

$$\mathbf{f}(\hat{\mathbf{u}}) + \mathbf{f}'(\hat{\mathbf{u}})\delta\hat{\mathbf{u}} = 0$$

i.e.

$$\delta\hat{\mathbf{u}} = -[\mathbf{f}'(\hat{\mathbf{u}})]^{-1}\mathbf{f}(\hat{\mathbf{u}}) \tag{A2.9}$$

provided the Jacobian matrix is invertible. This gives the basic step for the Newton iteration:

$$\delta\mathbf{u}^{(k+1)} = -[\mathbf{f}'(\mathbf{u}^{(k)})]^{-1}\mathbf{f}(\mathbf{u}^{(k)}). \tag{A2.10}$$

Hence

$$\mathbf{u}^{(k+1)} = \mathbf{u}^{(k)} - [\mathbf{f}'(\mathbf{u}^{(k)})]^{-1}\mathbf{f}(\mathbf{u}^{(k)}) \tag{A2.11}$$

is the improved estimate. This iteration converges provided $\mathbf{u}^{(0)}$ is close enough to the exact solution and the Jacobian is continuous and invertible. The difficulties are (a) getting the initial $\mathbf{u}^{(0)}$ and (b) the expense of forming and solving eqn (A2.10). It may not be easy or even possible to obtain the Jacobian matrix and there are various ways of approximating it. See Conte and de Boor (1980) for a fairly elementary presentation and Dennis and Moré (1977) for a more advanced discussion on quasi-Newton methods.

Appendix 3   Noise problems

This is a summary of the noise problems mentioned in this book. 'Noise' here means spurious oscillations in space or time or erroneous solutions introduced by the numerical scheme. It is important to stress that they should always be investigated. The key things to check are the time step, the grid (element) size, the boundary conditions, and the progress of any iteration.

Time stepping

Spurious solutions can be introduced by multi-step schemes—keep to single-step schemes (section 3.3).

With the θ method scheme (section 3.4), the most accurate method is given by $\theta = 0.5$ (Crank–Nicolson, trapezium rule). This is liable to produce oscillations with too large Δt (Crank–Nicolson sawtooth effect). Reducing Δt controls this; taking $\theta > 0.5$ reduces the oscillation and $\theta = 1.0$ removes it but the result is less accurate. The box scheme (sections 3.9 and 3.10) is most accurate with $\theta = 0.5$ and stable for $\theta \geq 0.5$. $\theta > 0.5$ may be recommended for some physical conditions (section 4.4).

Grid size

A finite difference grid or finite elements too large can produce spatial oscillations where the solution should be monotonic. This may be related to the mesh Peclet number (sections 3.8 and 9.5).

With time-dependent problems results with coarse grids do not improve with run-in (sections 7.8 and 7.9).

With finite elements there may be spatial oscillations due to under-integration (section 6.6). The patch test (section 6.7) can be used to check this.

Boundary conditions

There may be a problem with a sudden change in a boundary condition (section 4.2).

It is easy to put in a known-flow boundary condition with a wrong sign—water piling up or disappearing incongruously at a boundary is a sign of this.

Other odd behaviour of a model near a boundary may indicate the boundary should be farther out (section 8.3) or the boundary condition is inappropriate.

ADI schemes used with irregular boundaries can give trouble which is difficult to circumvent (Weare, 1979).

Initial conditions

The model may have to be run in to some dynamic steady state so that initial noise has disappeared before the actual numerical experiments start (sections 4.3 and 7.8).

Iteration

Oscillation with iteration indicates that underrelaxation (damping) is needed (sections 5.5 and 7.6).

Mention must also be made here of the classic paper by Gresho and Lee (1979). They argue very strongly that upwind schemes may be dangerous because they suppress noise by introducing numerical diffusion. The solution may look better but the grid may be too coarse for the numerical results to be meaningful. Upwinding is only acceptable if the numerical diffusion is significantly less than the physical diffusion. Gresho and Lee also strongly advocate the use of a consistent mass matrix always with the finite element method as lumped (diagonal) mass matrices give misleading results.

References

Abbott, M. B. (1979) *Computational Hydraulics*, Pitman, London.

Abbott, M. B. and Ionescu, F. (1967) On the numerical computation of nearly horizontal flows, *Journal of Hydraulic Research*, **5**, no. 2, 97–118.

Abbott, M. B., Bathurst, J. C., Cunge, J. A., O'Connell, P. E., and Rasmussen, J. (1986) An introduction to the European Hydrological System—Système Hydrologique Européen, 'SHE', 2: structure of a physically-based, distributed modelling system, *Journal of Hydrology*, **87**, 61–77.

Ackers, P. (1958) Resistance of fluids flowing in pipes and channels, *Hydraulics Research Paper no. 1*, HMSO, London.

Adey, R. A. and Brebbia, C. A. (1973) Finite element solution for effluent dispersion, in *Numerical Methods in Fluid Dynamics*, ed. C. A. Brebbia and J. J. Connor, Pentech Press, London.

Arnold, P. D. (1989) A numerical model of overland flow, *M.Sc. Dissertation*, Department of Mathematics, University of Reading, UK.

ASCE Task Committee (1988) Turbulence modeling of surface flow and transport, Parts I–V. *Journal of Hydraulic Engineering*, **114**, 970–1073.

Bach, H. K., Brink, H., Olesen, K. W., and Havnø, K. (1989) Application of PC based models in river water quality modelling, in *Proceedings of International Conference on Hydraulic and Environmental Modelling, Bradford*, ed. R. A. Falconer, P. Goodwin, and R. G. S. Matthew, Gower Technical, Aldershot.

Barlow, J. (1976) Optimal stress locations in finite element models, *International Journal of Numerical Methods in Engineering*, **10**, 243–251.

Bathurst, J. C. and Purnama, A. (1991) Design and application of a sediment and contaminant transport modelling system, *International Association of Hydrological Sciences, Publication no. 203*.

Bear, J. (1979) *Hydraulics of Groundwater*, McGraw-Hill, New York.

Benque, J. P., Cunge, J. A., Feuillet, J., Hauguel, A., and Holly, F. M. (1982) A new method for tidal current computation, *Journal of the Waterway, Port, Coastal and Ocean Division, American Society of Civil Engineers*, **108**, 396–417.

Berkowitz, B., Bear, J., and Braester, C. (1988) Continuum models for contaminant transport in fractured porous formations, *Water Resources Research*, **24**, no. 8, 1225–1236.

Bettencourt, J. M., Zienkiewicz, O. C., and Cantin, G. (1981) Consistent use of finite elements in time and the performance of various recurrence schemes for the heat diffusion equation, *International Journal of Numerical Methods in Engineering*, **17**, 931–938.

Beven, K. A., Calver, A., and Morris, E. M. (1987) The Institute of Hydrology distributed model, Institute of Hydrology Report 98, Institute of Hydrology, Wallingford, Oxon. England OX10 8BB.

Buckingham, E. (1914) On physically similar systems; illustrations of the use of dimensional equations, *Physical Review*, **4**, 345–376.

Buckingham, E. (1915) Model experiments and the forms of empirical equations, *Transactions of the American Society of Mechanical Engineers*, **37**, 263–296.

Bulirsch, R. and Stoer, J. (1966) Numerical treatment of ordinary differential equations by extrapolation methods, *Numerische Mathematik*, **8**, 1–13.

Burden, R. L., Faires, J. D., and Reynolds, A. C. (1981) *Numerical Analysis*, Prindle, Weber and Schmidt, Boston, MA.

Burns, A. D. and Wilkes, N. S. (1987) A finite difference method for the computation of fluid flows in complex three dimensional geometries, *UKAEA Harwell Laboratory* AERE R 12342, HMSO, London.

Calver, A. and Wood, W. L. (1989) On the discretization and cost-effectiveness of a finite element solution for hillslope subsurface flow, *Journal of Hydrology*, **110**, 165–179.

Calver, A. and Wood, W. L. (1991) Dimensionless hillslope hydrology, *Proceedings of the Institution of Civil Engineers, Part 2*, **91**, 593–602.

Chaudhry, Y. M. and Contractor, D. N. (1973) Application of the implicit method to surges in open channels, *Water Resources Research*, **9**, no. 6, 1605–1612.

Cheng, R. T., Casulli, V., and Milford, S. N. (1984) Eulerian–Lagrangian solution of the convection–dispersion equation in natural coordinates, *Water Resources Research*, **20**, (7), 944–952.

Clapp, R. B. and Hornberger, G. M. (1978) Empirical equations for some soil hydraulic properties, *Water Resources Research*, **14**, 601–604.

Connorton, B. J. (1980) Numerical solution of the quasi-linear regional groundwater equations with reference to the chalk aquifer of the Berkshire Downs, *M.Sc. Dissertation*, University of Reading.

Connorton, B. J. and Hanson, C. A. (1978) Regional modelling—analogue and digital approaches. Institution of Civil Engineers, Thames Groundwater Scheme Conference, Reading University, 61–76.

Connorton, B. J. and Reed, R. N. (1978) A numerical model for the prediction of long term well yield in an unconfined chalk aquifer, *Quarterly Journal of Engineering Geology*, **11**, 127–138.

Connorton, B. J. and van Beesten, D. P. (1983) *Lower Colne Gravels—Final Report*, Thames Water, Reading, Berks.

Connorton, B. J. and Wood, W. L. (1983) Noise problems with ephemeral streams in aquifer models, *International Journal for Numerical Methods in Fluids*, **3**, 201–208.

Conte, S. D. and de Boor, C. (1980) *Elementary Numerical Analysis*, McGraw-Hill, New York.

Cooley, R. L. (1983) Some new procedures for numerical solution of variably saturated flow problems. *Water Resources Research*, **19**, 1271–1285.

Courant, R., Friedrichs, K., and Lewy, H. (1967) On partial differential equations of mathematical physics, *IBM Journal*, **11**, 215–234 (English translation of (1928) 'Über die partiellen Differenzengleichungen der mathematischen Physik', *Mathematische Annalen*, **100**, 32–74).

Crank, J. (1975) *The Mathematics of Diffusion*, 2nd edition, Oxford University Press.

Crank, J. and Nicolson, P. (1949) A practical method for the numerical integration of solutions of partial differential equations of heat conduction type, *Proceedings of the Cambridge Philosophical Society*, **43**, 50–67.

Cunge, J. (1989) Review of recent developments in river modelling, in *Proceedings of International Conference on Hydraulic and Environmental Modelling, Bradford*, ed. R. A. Falconer, P. Goodwin, and R. G. S. Matthew, Gower Technical, Aldershot.

Cunge, J. A., Holly, F. M., Jr, and Verwey, A. (1980) *Practical Aspects of Computational River Hydraulics*, Pitman, London.

Darcy, H. (1856) *Les fontaines publiques de la ville de Dijon*, Dalmont, Paris.

Dennis, J. E., Jr and Moré, J. J. (1977) Quasi-Newton methods, motivation and theory, *SIAM Review*, **19**, 46–89.

De Smedt, F. (1990) Three-dimensional modeling of the subsurface transport of heavy metals released from a sludge disposal, in *Computational Methods in Subsurface Hydrology*, ed. G. Gambolati, A. Rinaldo, C. A. Brebbia, W. G. Gray and G. F. Pinder, Computational Mechanics Publications and Springer, Southampton.

Duff, I. S., Erisman, A. M., and Reid, J. K. (1989) *Direct Methods for Sparse Matrices*, Oxford University Press.

Dupuit, J. (1863) *Etudes théoriques et pratiques sur le mouvement des eaux dans les canaux et à travers les terrains perméables*, 2nd edition, Dunod, Paris.

Ekebjaerg, L. and Justesen, P. (1981) An explicit scheme for advection–diffusion modelling in two dimensions. *Computer Methods in Applied Mechanics and Engineering*, **88**, 287–297.

Ellis, J. and Pender, G. (1982) Chute spillway design calculations, *Proceedings of the Institution of Civil Engineers*, **73**, 299–312.

Elsworth, D. (1987) A boundary element–finite element procedure for porous and fractured media flow, *Water Resources Research*, **23**, part 4, 551–560.

Ergatoudis, J. G., Irons, B. M., and Zienkiewicz, O. C. (1968) Curved isoparametric quadrilateral elements for finite element analysis. *International Journal of Solids and Structures*, **4**, 31–42.

Evans, E. P. (1977) The behaviour of a mathematical model of open channel flow, in *Proceedings of the 17th Congress of the International Association for Hydraulic Research, Baden-Baden*, paper A97.

Evans, E. P. and Whitlow, C. D. (1991) Recent developments in one dimensional modelling of open channel networks, *Seminar on River and Flood Plain Management*, Scottish Hydraulics Study Group, c/o Department of Civil Engineering, Glasgow University.

Falconer, R. A. (1986) A two-dimensional model study of the nitrate levels in an inland natural basin, *International Conference on Water Quality Modelling in the Inland Natural Environment, Bournemouth*, England, Gower Technical, Aldershot, pp. 325–344.

Falconer, R. A. and Chen, Y. (1991) An improved representation of flooding and drying and wind stress effects in a two-dimensional tidal numerical model, *Proceedings of the Institution of Civil Engineers*, **91**, 659–678.

Ferraro, V. C. A. (1962) *Electromagnetic Theory*, The Athlone Press, London.

French, R. H. (1987) *Open-Channel Hydraulics*, McGraw-Hill, Singapore.

Frind, E. O. (1980) Seawater intrusion in continuous coastal aquifer-aquitard systems, in *Finite Elements in Water Resources* (ed. S. Y. Wang, C. A. Brebbia, C. V. Alonso, W. G. Gray and G. F. Pinder) University of Mississippi Press.

Frind, E. O. and Verge, M. J. (1978) Three-dimensional modeling of groundwater flow systems, *Water Resources Research*, 14, 844–855.

Gambolati, G., Galeati, G., and Neuman, S. P. (1990) A Eulerian–Lagrangian finite element model for coupled groundwater transport, in *Computational Methods in Subsurface Hydrology*, ed. G. Gambolati, A. Rinaldo, C. A. Brebbia, W. G. Gray, and G. F. Pinder, Computational Mechanics Publications and Springer, Southampton.

Gambolati, G., Rinaldo, A. Brebbia, C. A., Gray, W. G., and Pinder, G. F. (eds) (1990) *Computational Methods in Surface Hydrology*, Computational Mechanics Publications and Springer, Southampton.

Gibbs, N. E., Poole, W. G., and Stockmeyer, P. K. (1976) An algorithm for reducing the bandwidth and profile of a sparse matrix, *SIAM Journal of Numerical Analysis*, 13, 236–250.

Golub, G. H. and van Loan, C. F. (1990) *Matrix Computations*, Johns Hopkins University Press.

Goodwin, P., Matthew, R. S. G., and Wright, N. G. (1989) Prediction of flood wave due to dam failure using the Preissmann scheme, in *Proceedings of International Conference on Hydraulic and Environmental Modelling, Bradford*, ed. R. A. Falconer, P. Goodwin, and R. G. S. Matthew, Gower Technical, Aldershot.

Gragg, W. B. (1965) On extrapolation algorithms for ordinary initial-value problems, *SIAM Journal of Numerical Analysis*, 2, 384–403.

Gresho, P. M. and Lee, R. L. (1979) Don't suppress the wiggles—they're telling you something!, *Finite Element Methods for Convection Dominated Flow*, ed. T. J. R. Hughes, American Society of Mechanical Engineers, AMD34, 37–61.

Griffiths, A. L. (1978) A finite element model of groundwater flow in unconfined aquifers, *M.Sc. Dissertation*, University of Reading.

Grilli, S. (1989) Modelling of some elliptic fluid mechanics problems by the boundary element method, *Advances in Water Resources*, 12, 66–72.

Hawken, D. M., Townsend, P., and Webster, M. F. (1991) A comparison of gradient recovery methods in finite element calculations, *Communications in Applied Numerical Methods*, 7, 195–204.

Henry, H. R. (1959) Salt intrusion into fresh water aquifers, *Journal of Geophysics Research*, 64, (11), 1911–1919.

Henry, H. R. (1964) Effects of dispersion on salt encroachment in coastal aquifers, *United States Geological Survey Water Supply*, paper 1613C, C70–C84.

Hooper, A. G., Mitchell, G. M., and Powell, S. M. (1989) Implementation of water quality model for planning and design applications, in *Proceedings of International Conference on Hydraulic and Environmental Modelling, Bradford*, ed. R. A. Falconer, P. Goodwin, and R. G. S. Matthew, Gower Technical, Aldershot.

Hughes, T. J. R. and Brooks, A. (1979) A multi-dimensional upwind scheme with no crosswind diffusion, in *Finite Element Methods for Convection Dominated Flow* (ed. T. J. R. Hughes) American Society of Mechanical Engineers, AMD34, 19–34.

Hull, T. E., Enright, W. H., Fellen, B. M., and Sedgwick, A. E. (1971) Comparing numerical methods for ordinary differential equations. *University of Toronto Department of Computer Science Technical Report no.* 29.

Hutson, J. L. and Wagenet, R. J. (1991) Simulating nitrogen dynamics in soils using a deterministic model, *Soil Use and Management*, **7**, 74–78.

Huyakorn, P. S. and Nilkuha, K. (1979) Solution of transient transport equation using an upstream finite element scheme, *Applied Mathematical Modelling*, **1**, 187–195.

Huyakorn, P. S. and Taylor, C. (1976) Finite element model for coupled groundwater flow and convective dispersion, in *Proceedings of the First International Conference on Finite Elements in Water Resources*, Princeton University, Princeton, NJ.

Huyakorn, P. S., Mercer, J. W., and Ward, D. S. (1985) Finite element matrix and mass balance computational schemes for transport in variably saturated porous media, *Water Resources Research*, **21**, 346–358.

Huyakorn, P. S., Springer, E. P., Guvanasen, V., and Wadsworth, T. D. (1986) A three-dimensional finite element model for simulating water flow in variably saturated porous media, *Water Resources Research*, **22**, 1790–1808.

Jacoby, P. (1982) A mathematical model of groundwater flow in the Lower Colne Gravel aquifer, Department of Mathematics, University of Reading, *Report* for Thames Water, Reading, Berks.

Jensen, O. K. and Finlayson, B. A. (1980) Oscillation limits for weighted residual methods applied to convective diffusion equations, *International Journal of Numerical Methods in Engineering*, **15**, 1681–1689.

Johnsson, H., Bergström, L., Jansson, P. E., and Paustian, K. (1987) Simulated nitrogen dynamics and losses in layered agricultural soil, *Agricultural, Ecosystems and Environment*, **18**, 333–356.

Kaluarachchi, J. J. and Parker, J. C. (1988) Finite element model of nitrogen species transformation and transport in the unsaturated zone, *Journal of Hydrology*, **103**, 249–274.

Kuiper, L. K. (1987) A comparison of iterative methods as applied to the solution of the nonlinear three-dimensional groundwater flow equation. *SIAM Journal on Scientific and Statistical Computing*, **8**, 521–528.

Lambert, J. D. (1973) *Computational Methods in Ordinary Differential Equations*, Wiley, London.

Latinopoulos, P. (1986) A boundary element approach for modelling groundwater movement, *Advances in Water Resources*, **9**, 171–177.

Latinopoulos, P. and Katsifarakis, K. (1991) A boundary element and particle tracking model for advective transport in zoned aquifers, *Journal of Hydrology*, **124**, 159–176.

Lees, M. (1966) A linear three-level difference scheme for quasi-linear parabolic equations, *Mathematics of Computation*, **20**, 516–622.

Lencastre, A. (1987) *Handbook of Hydraulic Engineering, English Language Edition*, Ellis Horwood, Chichester.

Leonard, B. P. (1979) A stable and accurate convective modelling procedure based on quadratic upstream interpolation, *Computer Methods in Applied Mechanics and Engineering*, **13**, 59–98.

Lesaint, P. and Zlamal, M. (1979) Superconvergence of the gradient of finite element solutions. R.A.I.R.O. (Revue Francaise d'Automatique d'Informatique et de Recherche Operationelle) *Numerical Analysis*, **13**, 139–166.

Levine, N. (1985) Superconvergent estimation of the gradients from linear finite element approximation of triangular elements, *Ph.D. Thesis*, University of Reading.

Liggett, J. A. and Liu, P. L.-F. (1983) *The Boundary Integral Equation Method for Porous Media Flow*, Allen & Unwin, London.

Lui Sui-Qing and Falconer, R. A. (1989) Application of the QUICK difference scheme for two-dimensional water quality modelling, in *Proceedings of International Conference on Hydraulic and Environmental Modelling, Bradford*, ed. R. A. Falconer, P. Goodwin, and R. G. S. Matthew, Gower Technical, Aldershot.

Lynch, D. R. (1984) Mass conservation in finite element groundwater models. *Advances in Water Resources*, **7**, 67–75.

MacNeal, R. H. (1953) An asymmetrical finite difference network, *Quarterly Journal of Applied Mathematics*, **11**, 295–310.

Mangold, D. C. and Tsang, C-F. (1991) A summary of subsurface hydrological and hydrochemical models, *Reviews in Geophysics*, **29**, 51–79.

Mariño, M. A. (1981) Analysis of the transient movement of water and solutes in stream–aquifer systems, *Journal of Hydrology*, **49**, 1–17.

Matthews, R. S. and Wood, W. L. (1987) A cost-efficiency analysis of boundary flow calculations from a finite element solution to groundwater flow problems, *Communications in Applied Numerical Methods*, **3**, 145–154.

Morris, E. M. and Woolhiser, D. A. (1980) Unsteady one-dimensional flow over a plane: partial equilibrium and recession hydrographs, *Water Resources Research*, **16**, 355–360.

Narasimhan, T. N. and Witherspoon, P. A. (1976) An integrated finite difference method for analyzing fluid flow in porous media, *Water Resources Research*, **12**, 57–64.

Nawalany, M. (1990) Regional versus local computations of groundwater flow, in *Computational Methods in Subsurface Hydrology*, ed. G. Gambolati, A. Rinaldo, C. A. Brebbia, W. G. Gray, and G. F. Pinder, Computational Mechanics Publications and Springer, Southampton.

Neat, J. D., Jackson, L. D., and Falconer, R. A. (1989) Mathematical modelling of the rivers Aire and Calder, in *Proceedings of International Conference on Hydraulic and Environmental Modelling, Bradford*, ed. R. A. Falconer, P. Goodwin, and R. G. S. Matthew, Gower Technical, Aldershot.

Neuman, S. P. (1984) Adaptive Euler–Lagrange finite element method for advection dispersion, *International Journal for Numerical Methods in Engineering*, **20**, 321–337.

Nishi, T. S., Bruch, J. C., and Lewis, R. W. (1976) Movement of pollutants in a two-dimensional seepage flowfield, *Journal of Hydrology*, **31**, 307–321.

Osment, J., Reeve, D. E., Maiz, N. B., and Moussa, M. (1991) *Techniques for*

Environmentally Sound Water Resources Development, ed. R. Wooldridge, Pentech Press, London.

Paniconi, C., Aldama, A. A., and Wood, E. F. (1990) Time-discretization strategies for the numerical solution of Richards' equation, in *Computational Methods in Subsurface Hydrology*, ed. G. Gambolati, A. Rinaldo, C. A. Brebbia, W. G. Gray, and G. F. Pinder, Computational Mechanics Publications and Springer, Southampton.

Pender, G. (1992) Maintaining numerical stability of flood plain calculations by time increment splitting, *Proceedings of the Institution of Civil Engineers, Water, Maritime and Energy*, **96**, 35–42.

Pini, G., Gambolati, G., and Galeati, G. (1989) 3-D finite element transport models by upwind preconditioned conjugate gradients, *Advances in Water Resources*, **12**, 54–58.

Porter, J. D. and Jackson, C. P. (1990) Application of quasi-Newton methods to nonlinear groundwater flow problems, in *Computational Methods in Subsurface Hydrology*, ed. G. Gambolati, A. Rinaldo, C. A. Brebbia, W. G. Gray, and G. F. Pinder, Computational Mechanics Publications and Springer, Southampton.

Powell, M. J. D. (1970) A Fortran subroutine for solving systems of nonlinear algebraic equations, in *Numerical Methods for Nonlinear Algebraic Equations*, ed. P. Rabinowitz, Gordon and Breach, New York.

Preissmann, A. (1961) Propagation des intumescences dans les canaux et rivières, *First Congress of the French Association for Computation, Grenoble*.

Price, R. K. (1974) Comparison of four numerical methods for flood routing, *American Society of Civil Engineers, Journal of Hydraulics Division*, **100**, no. HY7, 879–899.

Reeve, D. E. and Hiley, R. A. (1992) Numerical prediction of tidal flow in shallow water, *International Conference on Coastal Engineering, Southampton*.

Refsgaard, A. and Jørgensen, G. H. (1990) Use of three-dimensional modelling in groundwater management and protection, in *Computational Methods in Subsurface Hydrology*, ed. G. Gambolati, A. Rinaldo, C. A. Brebbia, W. G. Gray, and G. F. Pinder, Computational Mechanics Publications and Springer, Southampton.

Richards, L. A. (1931) Capillary conduction of liquids through porous mediums, *Physics*, **1**, 318–33.

Richardson, L. F. (1910) The approximate arithmetical solution by finite differences of physical problems involving differential equations, with an application to stresses in a masonry dam. *Philosophical Transactions (A)*, **210**, 307–357.

Richardson, L. F. (1927) The deferred approach to the limit, I—single lattice, *Transactions of the Royal Society of London*, 226, 299–349.

Rubin, J. (1983) Transport of reacting solutes in porous media: relation between mathematical nature of problem formulation and chemical nature of reactions, *Water Resources Research*, **19**, 1231–1252.

Rushton, K. R. (1981) Modelling groundwater systems, in *Case Studies in Groundwater Resources Evaluation*, ed. J. W. Lloyd, Clarendon Press, Oxford.

Rushton, K. R. and Wedderburn, L. A. (1973) Starting conditions for aquifer simulations, *Ground Water*, **11**, 37–42.

Samuels, P. G. (1977) Efficiency of numerical methods, *Report no.* IT 163, Hydraulics Research Station, Wallingford, England.

Samuels, P. G. (1985) Modelling open channel flow using Preissmann's scheme, *2nd International Conference on the Hydraulics of Floods and Flood Control*, British Hydraulics Research Association, Cranfield.

Samuels, P. G. (1990) Cross-section location in one-dimensional models, *International Conference on River Flood Hydraulics*, Hydraulics Research, Wallingford, England, pp. 339–350.

Samuels, P. G. and Skeels, C. P. (1990) Stability limits for Preissmann's scheme, *American Society of Civil Engineers, Journal of Hydraulic Engineering*, **116**, 997–1012.

Segol, G., Pinder, G. F. and Gray, W. G. (1975) A Galerkin finite element technique for calculating the transient position of the saltwater front, *Water Resources Research*, **11**, 343–347.

Shapiro, A. M. and Andersson, J. (1983) Steady state response in fractured rock. A boundary element solution for a coupled discrete fracture continuum model, *Water Resources Research*, **19**, part 4, 959–969.

Silin-Bekchurin, A. I. (1958) *Dynamics of Groundwater* (in Russian), Moscow Izdat, Moscow University.

Smith, R. W. (1989) Review of recent developments in mixing and dispersion. International Conference on Hydraulic and Environmental Modelling, Bradford, ed. R. A. Falconer, P. Goodwin and R. S. G. Matthew, Gower Technical, Aldershot.

Sobey, R. J. (1984) Numerical alternatives in transient stream response, *Journal of Hydraulic Engineering*, **110**, 749–772.

Spencer, A. J. M., Parker, D. F., Berry, D. S., England, A. H., Faulkner, T. R., Green, W. A., Holden, J. T., Middleton, D., and Rogers, T. G. (1977) *Engineering Mathematics*, 2 vols, Van Nostrand Reinhold, Wokingham.

Spink, A. E. F. and Wilson, E. E. M. (1989) Groundwater resource management in coastal aquifers, *Groundwater Monitoring and Management* (*Proceedings of Dresden Symposium*), ed. T. E. O'Donnell, International Association of Hydrological Sciences, Delft Publication no. 173.

Sposito, G., Jury, W. A., and Gupta, V. K. (1986) Fundamental problems in the stochastic convection–dispersion model of solute transport in aquifers and field soils, *Water Resources Research*, **22**, 77–88.

Storm, B. (1991) Modeling of saturated flow and the coupling of surface and subsurface flow. Chapter 10 in Bowles, D. S. and O'Connell, P. E. (eds) *Recent Advances in Modeling of Hydrologic Systems*, pp. 185–203, Kluwer Academic Publishers, Delft.

Streeter, V. L. and Wylie, E. B. (1975) *Fluid Mechanics*, McGraw-Hill, New York.

Strang, G. and Fix, G. J. (1973) An analysis of the finite element method. Prentice-Hall, Englewood Cliffs.

Taylor, C. and Huyakorn, P. S. (1976) Finite element analysis of three-dimensional groundwater flow with convective dispersion, in *Finite Elements in Fluids*, Vol. 3, ed. R. H. Gallagher, O. C. Zienkiewicz, J. T. Oden, M. Morandi Cecchi, and C. Taylor, Wiley, Chichester.

Thacker, W. C. (1980) A brief review of techniques for generating irregular computational grids, *International Journal for Numerical Methods in Engineering*, **15**, 1335–1341.

Thompson, J. F., Warsi, Z. U., and Mastin, C. W. (1982) Boundary-fitted coordinate systems for numerical solution of partial differential equations, *Journal of Computational Physics*, **47**, 1–108.

Trkov, A. and Wood, W. L. (1980) Comparison between a finite element and a composite method for the numerical solution of a three-dimensional problem, *International Journal for Numerical Methods in Engineering*, **15**, 1083–1094.

van Genuchten, M. Th. and Jury, W. A. (1987) Progress in unsaturated flow and transport modeling, *Reviews of Geophysics*, **25**, 135–140.

Varga, R. S. (1962) *Matrix Iterative Analysis*, Prentice-Hall, Englewood Cliffs.

Vested, H. J., Justesen, P., and Ekebjaerg, L. (1992) Advection-dispersion modelling in three dimensions. *Applied Mathematical Modelling*, **16**, 506–19.

Wark, J. B., Samuels, P. G., and Ervine, D. A. (1990) A practical method of estimating velocity and discharge in compound channels, in *Proceedings of International Conference on River Flood Hydraulics*, Hydraulics Research Ltd., Wallingford, ed. W. R. White, Wiley.

Weare, T. J. (1979) Errors arising from irregular boundaries in ADI solutions of the shallow water equations, *International Journal for Numerical Methods in Engineering*, **14**, 921–931.

Wikramaratna, R. S. (1981) The flow of groundwater in a coastal aquifer. M.Sc. dissertation, University of Reading.

Wikramaratna, R. S. and Reeve, C. E. (1986) A model to simulate the effects of abstraction on a saline interface in a porous medium, in *Proceedings of the 9th Salt Water Intrusion Meeting*, Delft, The Netherlands, pp. 515–531.

Wikramaratna, R. S. and Wood, W. L. (1983) Control of spurious oscillations in the salt water intrusion problem, *International Journal for Numerical Methods in Engineering*, **19**, 1243–1251.

Wood, W. L. (1976) On the finite element solution of an exterior boundary value problem, *International Journal for Numerical Methods in Engineering*, **10**, 885–891.

Wood, W. L. (1990) *Practical Time-stepping Schemes*, Oxford University Press.

Wood, W. L. and Calver, A. (1990) Lumped versus distributed mass matrices in the finite element solution of subsurface flow, *Water Resources Research*, **26**, 819–825.

Wood, W. L. and Calver, A. (1992) Initial conditions for hillslope hydrology modelling, *Journal of Hydrology*, **130**, 379–397.

Wood, W. L. and Lewis, R. W. (1975) A comparison of time-marching schemes for the transient heat conduction equation, *International Journal of Numerical Methods in Engineering*, **9**, 679–689.

Woolhiser, D. A. and Liggett, J. A. (1967) Unsteady, one-dimensional flow over a plane—the rising hydrograph, *Water Resources Research*, **3**, 753–771.

Yeh, G. T. (1981) On the computation of Darcian velocity and mass balance in the finite element modelling of ground water flow, *Water Resources Research*, **17**, 1529–1534.

Zienkiewicz, O. C. and Taylor, R. L. (1989) *The Finite Element Method* (4th Edition) McGraw-Hill, Maidenhead.

Zlamal, M. (1977) Some superconvergence results in finite element methods, in *Mathematical Aspects of Finite Element Methods*, Springer, Berlin, pp. 353–362.

Author index

Numbers in bold type refer to the references

Subject index